THE INTERNET AND TELECOMMUNICATIONS POLICY

Selected Papers from the

1995 Telecommunications

Policy Research Conference

D1456839

TELECOMMUNICATIONS
A Series of Volumes Edited
by Christopher H. Sterling

THE INTERNET AND TELECOMMUNICATIONS POLICY

Selected Papers from the

1995 Telecommunications

Policy Research Conference

Gerald W. Brock
George Washington University

Gregory L. Rosston
Federal Communications Commission

  LAWRENCE ERLBAUM ASSOCIATES, PUBLISHERS
1996   Mahwah, New Jersey

Lawrence Erlbaum Associates, Inc., Publishers
10 Industrial Avenue
Mahwah, New Jersey 07430-2262

Cover design by Jennifer Sterling

Library of Congress Cataloging-in-Publication Data

Telecommunications Policy Research Conference (23rd : 1995)
The internet and telecommunications policy : selected papers from
the 1995 Telecommunications Policy Research Conference / edited by
Gerald W. Brock, Gregory L. Rosston.
    p. cm.
Includes bibliographical references and index.
  ISBN 0-8058-2418-9 (cloth: alk. paper). — ISBN 0-8058-2419-7
(pbk.: alk. paper).
  1. Telecommunication policy—United States—Congresses.
2. Information superhighway—United States—Congresses.
I. Brock, Gerald W. II. Rosston, Gregory L. III. Title.
HE7781. T44 1995
384'.068—dc20   96-23166
                                        CIP

Books published by Lawrence Erlbaum Associates are printed on acid-free
paper, and their bindings are chosen for strength and durability.

Printed in the United States of America
10 9 8 7 6 5 4 3 2 1

Contents

Authors

Timothy J. Brennan
University of Maryland

Gerald W. Brock
George Washington University

David Clark
MIT

Nicholas Economides
New York University

Charles Eldering
General Instrument Corporation

Deborah Estrin
University of Southern California

Joseph M. Foley
The Ohio State University

Shai Herzog
University of Southern California

Nageen Himayat
General Instrument Corporation

William H. Lehr
Columbia University

Jessica Litman
Wayne State University

Jeffrey MacKie-Mason
University of Michigan

Roger G. Noll
*The Brookings Institution
  and Stanford University*

Nosa Omoigui
Microsoft Corporation

Jon M. Peha
Carnegie Mellon University

David P. Reed
Cable Television Laboratories, Inc.

Stephen F. Roehrig
Carnegie Mellon University

Gregory L. Rosston
Federal Communications Commission

Scott Shenker
Xerox Palo Alto Research Center

Marvin A. Sirbu
Carnegie Mellon University

Richard Stannard
*New York State Public Service
  Commission*

Hal R. Varian
University of California, Berkeley

Qiong Wang
Carnegie Mellon University

Christopher Weare
University of Southern California

Martin B.H. Weiss
University of Pittsburgh

Lawrence J. White
New York University

Yuehong Yuan
Carnegie Mellon University

Acknowledgments

For many years, the Telecommunications Policy Research Conferences (TPRCs) have produced high quality research and commentary on a wide range of telecommunication policy issues. Although the conference papers have resulted in numerous publications in many different journals, there has not been a routine publication channel for the research presented at the conference. Through the efforts of Christopher Sterling, the series editor of the LEA Telecommunications Series, Hollis Heimbouch of LEA, and John Haring and Bridger Mitchell in their roles as former chairpersons of the TPRC Board of Directors, an arrangement was made for LEA to publish a series of volumes based on research presented at the TPRC. This is the second volume in that series and consists of revised versions of selected papers presented at the 23rd annual TPRC, held in October 1995.

I am grateful to the many members of the organizing committee and session moderators who served as referees to help choose the highest quality papers for inclusion in this volume.

The 23rd annual TPRC, and therefore the publication of this volume, was made possible through a generous grant from the John and Mary R. Markle Foundation.

Chapters 2 and 4 were first published in the *Antitrust Bulletin*. Chapter 15 was first published in the *Oregon Law Review*.

I am grateful for the exceptional help that Karoline Martin has provided in coordinating the many tasks necessary to produce both this volume and last year's TPRC volume.

Gerald W. Brock

Foreword

David P. Reed
Cable Television Laboratories, Inc.

Welcome to the second printed edition of selected papers from the Telecommunications Policy Research Conference (TPRC). As the Chair of the organizing committee for this year's conference, it is my honor to introduce the selected works of the 23rd annual TPRC.

TPRC has become the pre-eminent conference reporting on research of telecommunications policy issues. While the conference is now a respectful 23 years old, this is only the second printed edition of selected papers from the conference. One can sense a new tradition of scholarship associated with TPRC is developing with the annual publication of this book series.

At the outset I want to thank and congratulate the members of the organizing committee for their contributions in leading the conference. These members include Marjory Blumenthal of the National Research Council, Tim Brennan of the University of Maryland, Martin Cave of Brunel University, Allan Hammond of New York Law School, David Lytel of the Office of Science and Technology Policy of the White House, Russ Neuman of Tufts University, Lisa Rosenblum of the New York Public Service Commission, and David Waterman of Indiana University. Without their efforts the quality of the conference that attendees traditionally expect would not have been achieved this year.

Given the large number of changes that occurred in the fast-moving telecommunications policy field in 1995, the organizing committee was faced with numerous difficult decisions in reaching the final composition of the program. In the end, the program reflected the continuing development of several themes, although two themes stand out in importance given the events of 1995.

The first such major theme looked at the continued struggle of lawmakers and regulators at both the federal and state levels to further open local telecommunications markets to competition. One TPRC plenary session therefore examined elements of the proposed telecommunications reform bill slated for passage in

Congress. In 1995 several states took the lead as well by providing case studies of different regulatory models for more competitive local telephone markets. A report on the preliminary results of one such trial in Rochester, New York, provided one of the most memorable opening TPRC plenary sessions in recent memory.

A second major theme of the TPRC program examined the policy research questions raised by the explosive growth of the Internet and other data services for individuals and the commercial sector. Long a bastion of academic and government users, in 1995 consumers and businesses took to the new data services in unprecedented numbers. By devoting one plenary and other sessions to examine this development, TPRC'95 marked the arrival of these data services as an important new telecommunications policy research topic. It seems likely that future TPRC gatherings will feature a number of sessions dedicated to further exploration of these exciting new research issues.

To be sure, the program reflected other important policy research themes for 1995. Notable among these: the completion by the Federal Communications Commission (FCC) of the first spectrum auction of monetary significance through the tendering of personal communications services (PCS) licenses, and the completion by the FCC of its standard for advanced television, raising new interoperability and policy issues in the process.

The organizing committee is extremely grateful to all individuals who proposed papers for inclusion in the conference. With well over 100 paper proposals received from the Call for Papers, a large number of excellent papers could not be included given the short 2-day time span of the conference.

In addition, the organizing committee is indebted to the session chairs for their leadership in organizing and moderating their respective sessions. The conference leans heavily on these individuals to select the best papers within each topic and manage each session, and this group did an outstanding job.

Last, but not least, the organizing committee also appreciates the excellent administrative support provided by Dawn Higgins and Lori Rodriguez. More than any other individuals, they translated conference plans into reality, and did so in a first-rate fashion to the benefit of all. This book edition is yet another example of the benefits that flow from the excellent administrative foundation they represent and work hard to maintain.

On that closing note, I hope you enjoy this year's edition of selected works from the TPRC, and I look forward to your attendance at next year's conference.

Introduction

Gerald W. Brock
George Washington University

Gregory L. Rosston
Federal Communications Commission

After years of useful but obscure service, the Internet became an object of ubiquitous public discussion during 1995. The transformation of the Internet was symbolized by Mark Andreessen's spectacular ascent from a $6.85/hour student programmer at the University of Illinois to a 24-year-old cofounder of Netscape holding stock worth over $100 million. Numerous other instant fortunes from the stock market valuation of Internet-related companies indicate great public expectations for its future. The Internet represents a method of providing communications that is outside of the established structure of telecommunication policy. Established policy emphasizes clear boundaries between various segments of the communications industry. One set of policies applies to common carriers, such as telephone companies, while a different set of policies applies to broadcasters. The Internet mixes elements of the transmission (common carrier) concerns of the telephone industry and the content concerns of the broadcast industry with a perspective from the computer industry that regulation is not needed at all.

The Internet is of great importance: It represents a new integrated approach to the telecommunication industry that raises fundamental policy problems. Steady sustained technological advances in computers and electronics have caused a shift from traditional analog methods of providing communications to digital techniques with extensive computer intelligence. The Internet was created as a system of interconnected networks of communications lines and controlling computers. Formally, the Internet is an "enhanced service" under the Federal Communications Commission (FCC) categories and is exempted from regulation. As the unregulated Internet (including the World Wide Web and commercial online services) developed, it incorporated three different kinds of integration that raise new issues for telecommunication policy.

First, the Internet is created by the integration of multiple networks provided by independent entities with no overall control other than standards for interconnection protocols. The Internet represents the fullest expression to date of the unregulated "network of networks," which is widely expected as a model for future communications. Although the telephone industry has moved far from the old model of unified monopoly, it is still tightly controlled by regulation and central planning. Local exchange companies dominate their geographical areas. Long-distance companies and others connect to the local exchange companies with arrangements that are closely regulated. The assumption at the time of the 1984 divestiture of AT&T was that the local exchange companies would remain monopolies, and that there should be controls to guarantee access for competitive segments of the industry to the "bottleneck monopolies." Although some competition has come to the local exchange market place in the intervening years, the basic policy framework has remained concerned with hanging competitive elements onto a core monopoly structure. The future telecommunication network is likely to be made up of many different interconnected networks without any core monopoly as its anchor. In such a network, the past practice of rigidly structured "access charges"—based on arbitrary cost-allocation formulas—is no longer viable. A significant policy problem is creating interconnection arrangements among multiple competing networks that achieve efficiency and allow competition to flourish.

Second, the Internet includes the integration of multiple types of services with substantially different technical characteristics onto a single network. The Internet is used to transmit short e-mail messages, graphics, large data files, audio files, and (very slowly) video files. The various kinds of transmissions have vastly different bandwidth requirements and time sensitivities. A short video segment may require the transmission capacity of thousands of e-mail messages. A modest delay in transmitting packets may have no significant effect on e-mail or data file transfers while the same delay in transmitting audio or video may make the transmission worthless. In contrast to the integrated nature of services offered over the Internet, past telecommunication policy has assumed (and sometimes mandated) separate facilities and policies for different kinds of transmissions. The telephone network and associated policies are built around the switched two-way voice grade circuit, with an assumption that the predominant use of that circuit is to carry the human voice. The broadcast and cable-TV networks and the associated policies are based on providing one-way nonswitched transmission of video signals. Technological advances in fiber-optic transmission of signals and in compression of digital video signals have created the possibility of future integrated networks that carry all kinds of signals as digital packets of information, breaking down the policy boundaries of the past and creating the need for new integrated policies.

Third, the Internet and commercial networks connected to it include the integration of the provision of transmission capacity with various degrees of the provision of information. Past telecommunication policies have sharply distinguished between providers of communications capacity and providers of information content. Common carriers were required to transmit all information submitted to them in a nondiscriminatory way, and therefore had no editorial control over the information transmitted or any responsibility for that information. Broadcasters were required to operate "in the public interest" with regard to the material they transmitted. They have been responsible for that material and have been required to meet a varying set of standards for appropriate material over the years: indecency restrictions, public service and news programming, limitations on advertising time, children's programming requirements, and so forth. In the Internet, and in the expected future communications industry, those dividing lines are blurred as individual companies provide capacity to transmit communications for others and also provide their own content. For example, commercial online services provide transmission capacity (e.g., e-mail, Internet access) for their customers without regard to the information the customers are transmitting over that capacity. They also provide information over which the companies exercise editorial discretion. The fully integrated provision of content and transmission capacity raises new policy issues that have been illuminated in the debate regarding "indecent" material on the Internet and in the effort to define copyright standards for Internet material. Existing policy standards for dealing with content issues do not naturally translate into the categories needed for the Internet and other integrated providers of content and communications capacity.

The Telecommunications Act of 1996, signed by President Clinton on February 8, 1996, accelerated the process of integration in the communication industry that was already under way. The new law makes major revisions to the Communications Act of 1934 that will increase integration within the industry. The act eliminates the restrictions on activities of the Bell Operating Companies that were contained in the 1982 antitrust Consent Decree, allowing the Bell Operating Companies to enter the long-distance market. The act requires the removal of local barriers to entry, allowing long-distance companies and other potential entrants the ability to enter the local telephone market. It eliminates the statutory restrictions on the provision of cable-TV service by telephone companies, allowing them freedom to integrate the provision of telephone and video services over the same network. It also eliminates statutory restrictions limiting the ownership of cable-TV services by broadcasters, but does not require the elimination of related FCC regulatory restrictions.

The new law's primary effect is to break down barriers to entry into various segments of the telecommunications market. Past policy has contained a complex mixture of statutory law, FCC rules, state regulatory rules, and antitrust actions, which created walls around particular segments of the telecommunication industry. Companies operating in one segment were limited in their ability to enter other segments. That system of segmentation was already gradually breaking down prior to the passage of the new law. The new law condones recent efforts to reduce those barriers, and makes sweeping reductions in the barriers among industry segments that otherwise would have required years of gradual regulatory efforts to accomplish. Under the new law, it is likely that the entire telecommunications industry will move more toward the competitive network of integrated services that are already observed in the Internet. The industry of the future will probably be characterized by competing interconnected, vertically integrated companies providing a wide range of services with digital technology over a single set of physical facilities, along with numerous smaller companies providing services in specialized niches of the market.

Although the chapters in this volume were written prior to the passage of the new law, their importance is enhanced by the new law. These chapters do not deal with specific regulatory rules that are likely to be changed under the new law. Rather, they deal with the fundamental and complex policy problems that arise when previously separate segments of the telecommunications industry are integrated. The chapters were motivated by the observed integration of the Internet and other segments of the industry prior to the passage of the new law, and the expected increased integration of the future. They were designed to provide insights for the long-term development of policy. With the passage of the new law, the time frame for developing appropriate policies for an integrated industry has been shortened. Changes that were expected to occur over a period of several years will now occur much more rapidly. These chapters provide insights to guide the transition in the industry and the many specific policies that will need to be developed as the full implications of the new law are clarified.

Part I deals with problems of transforming local exchange telephone service from a monopoly in each geographical area to an interconnected competitive network of networks. Brennan (chap. 2) provides an explanation and analysis of the restrictions imposed on the Regional Bell Operating Companies (RBOCs) as part of the settlement of the government's antitrust case against AT&T. Brennan states that the various restrictions were generally based on the combination of two factors: market power (or control of "bottleneck" facilities) and regulation. Since the signing of the Modified Final Judgment (MFJ), the RBOCs have submitted numerous petitions to the court, as well as legislative proposals, to amend or eliminate the prohibitions imposed on them. Brennan's chapter analyzes the theory behind the original divestiture and the line of business restrictions looking at the two major ways in which the firms could misuse their position, discrimination and cross-subsidization or cost shifting. Using evidence from numerous submissions by the RBOCs regarding "changed circumstances," he analyzes whether the changes address four key questions: Have the incentives changed? Has the ability changed? Can regulators and antitrust authorities detect anticompetitive behavior?

Have efficiencies increased so that they outweigh potential harms? Although the Telecommunications Act of 1996 eliminated the prohibitions, the issues examined by Brennan remain relevant for future policy. The fundamental issue is the trade-off between full freedom for an incumbent company to benefit from potential economies of scope by providing a wide variety of services versus restrictions to prevent the anticompetitive use of an established market position. That trade-off will continue to be an issue as local competition is developed, and an understanding of the reasoning behind the MFJ restrictions will be important for understanding the new issues even if the solutions reached are different from the solution adopted in 1982.

One specific requirement of the new law is that local telephone companies make their facilities available at a wholesale rate to companies wishing to enter the local exchange market by reselling the incumbent's facilities. The initial development of long-distance competition was greatly enhanced by a resale requirement that allowed competitors to extend service beyond their own facilities by reselling capacity obtained from AT&T. It is expected that resale opportunities will be critical to the development of local exchange competition. The primary experience to date with resale of local exchange comes from the Rochester Telephone Corporation. Rochester filed its Open Market Plan in 1993, and created a wholesale entity that would deal with its own retail arm on the same terms and conditions as with resellers. Stannard (chap. 3) gives an interim assessment of the plan after 9 months from the perspective of a state regulator. He focuses on the effects of the wholesale discount, the quality of the network service, the flow of information, and service provided to competitors. This description is useful for competitors and regulators looking at how to implement resale competition in other geographical areas. Many of the competitive factors that have assumed paramount importance in Rochester were initially seen as insignificant. The Rochester experience points to the fact that implementation details are important for constructing a level playing field to promote competition.

Entrants to the local exchange market require interconnection with the incumbent carriers to exchange traffic and allow subscribers to both entrants and incumbents to reach all other subscribers. A critical factor in the viability of entrants is the terms and conditions of their interconnection agreements. This has been the focus of considerable debate around the world, wherever new entrants want to challenge the existing local exchange monopoly. In New Zealand, the efficient component pricing rule (ECPR) has been accepted by the courts (after several rounds of appeal) as an acceptable method for determining the price of interconnection. This principle states that the price charged for interconnection should be equal to the incremental cost of interconnection, including the incremental opportunity cost of the foregone sales (i.e., the profit that would have been realized from those retail sales that are diverted to the firm purchasing interconnection). Economides and White (chap. 4) take issue with some of the assumptions underlying the alleged efficient outcome this rule generates. They examine the impact of relaxing some of the assumptions required for the optimality of the ECPR. By relaxing these assumptions, they show that ECPR may inefficiently exclude new entry and eliminate the procompetitive benefits from additional entry. They examine the case where the monopolist initially sells the complementary product above marginal cost. Using Bertrand and Cournot models of competition, they show that even with "inefficient" competitors, social welfare can be increased through the price-reducing effects of competition. They also investigate the implication of relaxing other assumptions underlying ECPR—perfect substitution and no economies of scale in the complementary product—and show that use of ECPR can lead to a social welfare loss.

Part II considers the pricing problems that arise in an integrated network carrying traffic of different types across multiple service providers. The Internet is an example of such a network in use today. The provision of communication through multiple networks—each carrying multiple services with different technical characteristics—is expected to become routine in the future. Current Internet pricing to end users is generally very simple (such as a flat rate per month for unlimited usage, or a rate per hour) in contrast to the complex pricing plans proposed by these chapters. Part II illuminates the incentives and problems that arise as multiple kinds of traffic are mixed together.

Wang, Sirbu, and Peha (chap. 5) consider the impact of asynchronous transfer mode (ATM) technology on the network pricing problem within a single-provider network. They focus on the opportunity cost of adding new services to a network of fixed capacity. Their model looks at the pricing problem from a social welfare maximization perspective, and compares that with the results of a profit maximizing model. They conclude that the optimal pricing scheme incorporates the opportunity cost of using capacity, and also includes the opportunity cost of reserving capacity for guaranteed service.

Whereas Wang, Sirbu, and Peha examine the implications of pricing for different services within a single network, Lehr and Weiss (chap. 6) look at the pricing implications of homogeneous services provided by interconnected networks. Lehr and Weiss extend earlier work by Mackie-Mason and Varian to the multiple network context, and look at the implications for settlements between carriers. Although the routing of traffic may be complex, Lehr and Weiss show that it may be possible to set optimal congestion charges using only local information on costs and traffic. They conclude that usage pricing is unavoidable and desirable for the Internet, and that optimal pricing can be achieved with sharing of local congestion pricing information and a settlements process. However, they caution that the potential abuse of market power could significantly affect their results.

Shenker, Clark, Estrin, and Herzog (chap. 7) challenge the optimality paradigm that forms the basis for the conclusions in the prior two chapters for three reasons. First, marginal cost prices are unlikely to provide sufficient revenue to cover costs.

Second, it is difficult to obtain accurate real-time congestion costs because the actual path on a packet network is not known in advance. Finally, they posit structural goals other than optimality. Instead of marginal cost pricing, they propose "edge pricing," which is between usage-based pricing and flat pricing. Edge pricing incorporates the expected congestion along the expected path so that charges can be determined and assessed locally (at the "edge") where the packets begin their journey. Such pricing can take the form of flat-rate or usage-sensitive pricing. They conclude with a discussion of the infrastructure needed to support their pricing proposal.

Part III examines the problem of achieving interoperability in complex networks. Simply interconnecting networks physically is useless unless the software and standards used allow information initiated on one network to be understood by another network. There is both a technical and an economic side to achieving interoperability. As a technical matter, interoperability requires standards for data coding, digital compression, and so forth, along with methods to translate from one standard to another. As an economic matter, companies have varying incentives to achieve interoperability according to their market position, and therefore may have difficulty agreeing on technical requirements because some parties do not find interoperability in their interest.

Reed (chap. 8) examines the four layers that are used to describe digital television systems. He discusses the components and various technological choices that make up the presentation, compression, transport, and transmission layers, and provides information on the cost trade-offs for the different transmission media that are expected to carry digital video signals. He then provides an analysis of the expected degree of interoperability for each layer. Reed predicts that interoperability in the first generation of digital video will be limited to the presentation layer and may be extended to the compression and transport layers, but will not be available across transmission formats.

Weare (chap. 9) uses a transactions–cost approach to examine how the various corporate structures and market interactions affect interoperability in communications and computers. He identifies two key factors—asset specificity and uncertainty—as the driving forces behind firms' decisions whether to out source specific equipment. Higher asset specificity and uncertainty implies that a firm would be more susceptible to opportunistic behavior and less able to coordinate adaptation, and therefore more likely to internalize a particular function. He argues that vertical integration is a rational and efficient response to mitigate opportunistic behavior, and that policymakers should be tolerant of vertical integration. Weare examines possible strategic uses of interoperability, as well as the ability of regulatory authorities to detect anticompetitive behavior. He also distinguishes "easy" and "hard" cases. The hard cases present a difficult policy trade-off because a firm can minimize transactions costs while it minimizes competitors' ability to enter with compatible products. He concludes with applications of the framework to the government's antitrust investigation of Microsoft and the promotion of the Advanced Intelligent Network.

Omoigui, Sirbu, Eldering, and Himayat (chap. 10) compare the features of various network architectures for combining narrowband, broadband, and interactive services: hybrid fiber-coax, digital loop carrier, fiber to the curb, advanced hybrid fiber-coax, fiber to the curb with hybrid fiber-coax overlay, all fiber to the curb, and digital loop carrier with hybrid fiber-coax overlay. Using the features of each model, they present various scenarios of demand and model the cost of each architecture to develop cost comparisons for the various architectures. They conclude that the least-cost solution is to use hybrid fiber-coax to carry all services because the initial infrastructure costs are relatively low. However, if penetration rates rise above 50%, an all-fiber system may become the least-cost solution. They conclude that cable companies are well positioned to compete with telephone companies for communications delivery in the future.

Part IV considers issues of intellectual property that arise in expected integrated networks of the future. As information is increasingly stored in digital form and transmitted over networks that may provide integrated transmission and content services, established methods of dealing with social control of undesirable material, free speech rights, and protection of intellectual property rights will need substantial modification.

MacKie-Mason, Shenker, and Varian (chap. 11) analyze the relationship between network architecture and the nature and pricing of the content made available to consumers. In a single-provider architecture, they compare the content and pricing on application-blind networks, application-aware networks, and content-aware networks. They note that in an aware architecture, the operator has the ability to price differentiate and extract more of the surplus. Price differentiation leads to differences in the provision of content over the different types of networks. They examine the impact of different levels of awareness on welfare, liability, and incentives for content creation. They find that in a monopoly network, there may be more incentives for content creation with a blind network, but that under certain conditions welfare may be higher with an aware architecture than with a blind architecture.

Foley (chap. 12) examines in detail one of the aspects that MacKie-Mason et al. touch on—the potential liability for the network operator. With a legal, rather than an economic, framework, Foley looks at the history of free speech litigation and the rules placed on network operators to see how future restrictions may be upheld. He concludes that the courts are unlikely to uphold even content-neutral rules because they impose some limitation on free speech. Instead, he argues that the courts believe that regulators can construct other methods to achieve their goals without inhibiting free speech rights. However, Foley points out that the abolition of restrictions may be harmful to providers because of the potential for liability. He argues that some network providers would be willing to give up some of their newly acquired freedoms to protect themselves from liability for any transmissions they carry.

The huge subscription and storage cost for institutional libraries have led many of them to greet the digital age with enthusiasm because they expect to access electronic versions of scholarly journals more easily and at lower cost than printed journals. Noll (chap. 13) presents a model of scholarly publication that suggests electronic journals will not solve the library cost problem. Although each library is individually rational in choosing to adopt new technology and take advantage of the ease of "borrowing" articles from others, the collective actions are likely to increase costs for all, rather than decreasing costs. Noll develops a simple model of journal publication using first-copy costs and the marginal publication costs, and incorporates the history of journal proliferation. He then examines the effect of illegal duplication and the reduction in subscriptions on prices. Because a significant portion of costs are accounted for by the first-copy costs, subscription reductions are likely to lead to increased subscription prices. Because libraries are more likely to continue subscriptions, they will bear the burden of higher prices. In addition, the under provision of scholarly journals is likely to continue, and may even be exacerbated.

Yuan and Roehrig (chap. 14) look at the effect of the reduction of duplication and distribution costs on optimal copyright protection. They develop a monopolistic competition model of information products to examine the effects of increased competition, globalization of markets, and reduction in copying and distribution costs on copyright duration. They find that intensified competition should decrease the life of the copyright. According to their model, the optimal copyright moves from 45 years with a monopoly producer down to 16 years with a competitive environment. They find that decreased copying costs and globalization of markets should not affect the duration. They also use their model to examine the effects of product differentiation, economic life of the product, and returns to scale.

Litman (chap. 15) focuses on the applicability of copyright law in the digital age. Much attention has been paid to what changes, if any, are necessary for copyright law. Some argue that wholesale changes are needed because of the technology, whereas others argue that only fine tuning and clarification are necessary. Litman argues that the existing law is written to benefit those with entrenched interests, and that technological changes will fundamentally change the way in which copyrighted products are used. As a result, fundamental changes are needed in copyright law. She argues that the critical concern of copyright law should be the incentives to produce intellectual property, rather than prohibiting unauthorized copies. Therefore, she proposes to jettison the "copy"right law and move to a system based on commercial and noncommercial interests. Although this would not result in a bright line, she argues that it would align more closely with incentives, and would help correct abuses of the existing system.

I

LOCAL COMPETITION

Is the Theory Behind U.S. v AT&T Applicable Today?

Timothy J. Brennan
University of Maryland

INTRODUCTION

Ever since it was announced in 1982 and enacted in 1984, the Modified Final Judgment (MFJ) adopted to settle the Department of Justice's eight year antitrust suit against AT&T has been the subject of intense scrutiny and debate among lawyers, economists, policy makers, industry participants, and the public at large. Decisions by the Department of Justice and the courts narrowed the scope of these restrictions, only to intensify the decibel level from their proponents, who asserted greater vulnerability to assorted anticompetitive threats, and their opponents, primarily the regional Bell operating companies (RBOCs). Technological changes, primarily through the digitization of wireline networks, explosion in data communications, development of switched wireless systems, and plunging prices for complementary computers and multimedia equipment have made telephone industry even more prominent. The most prominent response is the recently passed Telecommunications Act of 1996, which vacated the MFJ and established conditions for the entry by the RBOCs into other markets.

Lost in the rush to reform is the idea that there were some reasons to force AT&T to divest its local telephone operations. These reasons may have been primarily theoretical, backed by interpretations of AT&T's pre-divestiture conduct rather than econometric evidence that keeping local telephone exchange companies from operating in other related markets would improve economic performance. One can debate whether these reasons caused the divestiture, or whether they served as an intellectual rationale for a political and legal decision. Nevertheless, unless one adopts a completely deterministic view of the policy process, there is some merit in asking the forgotten question: If the divestiture made sense in 1984, why would it not make sense in 1995?

The heated rhetoric surrounding this question is one compelling motivation for a systematic assessment of whether the "old" theory currently applies. Moreover, the conventional indicators of whether antitrust complaints are justified are mixed in this situation. The so-called First Theorem of Antitrust—"If a competitor complains about X, X must be good"—is not helpful. The telephone companies whose practices are currently circumscribed by the MFJ stand as both input suppliers and as potential competitors to the complaining firms.

It may be useful to look at the leading examples. The economist's instinct would be to treat AT&T's or MCI's opposition to allowing local telephone companies to provide interexchange service as yet another example of using antitrust, regulation, and the political process to attempt to insulate oneself from competition. A similar conclusion would follow from observing the opposition of the cable industry to moves to rescind legislative obstacles to telephone delivery of video services. However, both long distance telephone companies and cable companies currently rely on local telephone company services to be able to provide their services. Long distance companies usually require local telephone company connections providers to originate calls, and almost always need those connections to complete calls. Cable companies have traditionally used telephone company poles or conduits to spread their distribution and feeder cables out to reach homes. As such, these firms are, according to the "old" theory, as potentially vulnerable to the exercise of monopoly power by regulated local telephone companies *qua* input supplier in economically bad ways as they may be vulnerable to those companies *qua* competitor in good ways.

This review of the current debate begins with a restatement of that theory, laid out in detail in the pages of the *Antitrust Bulletin* around seven years ago.[1] Despite the attention given to the AT&T's divestiture of its local telephone operations and the concomitant line of business restrictions the MFJ placed on the divested holding companies, the mechanics of the theory, especially the role of regulation, are insufficiently appreciated. In reviewing the grounds for the divestiture, four crucial factors become apparent—the nature of regulation, the degree of market power, the effectiveness of injunctive relief, and the size of economies of scope. Assuming that these factors justified the divestiture in 1984—a perhaps controversial assumption, but one not currently under much dispute—the legal question will turn on whether circumstances regarding these factors have changed sufficiently in the last ten years to warrant effective repeal of the divestiture by allowing local telephone companies to reintegrate back into long distance service and to diversify into video programming.

---

1. Brennan, *Why Regulated Firms Should Be Kept Out of Unregulated Markets: Understanding the Divestiture in U.S. v. AT&T,* 32 ANTITRUST BULL. 747 (1987).

Fortunately, the policy debate has produced considerable economic documentation that addresses the "changed circumstances" question. Seeking vacation of the MFJ, some RBOCs sponsored a collection of forty-seven supporting affidavits, from fifty-six leading economists and telecommunications policy experts.[2] Examining those affidavits should provide some insight into the strength of the theoretical and empirical case for unfettering the RBOCs, and perhaps offer some advice as to the remaining question Congress, the FCC, and the courts need to answer before deciding whether and under what circumstances this unfettering should take place.

THE "OLD" THEORY, TO THE PRESENT DAY

Despite all of the academic, legal, and political attention devoted to *U.S. v. AT&T* and the MFJ, the economic rationale for the case remains widely misunderstood. There are two things the case was clearly *not* about, nor should have been. The first was that AT&T was an example of the proposition that size should be a *per se* antitrust violation. It is worth remembering that at the same 1982 press conference at which the Department of Justice announced the resolution of its eight year long case against AT&T via the divestiture of the local operating companies, it dropped its thirteen year old case against IBM, stating that the latter case was "without merit." A second candidate, recently appearing in press commentary inspired in part by the rise of Microsoft and the demise of IBM in the computer industry, is that some big firms would be better off leaner and meaner. Advocates of this rationale suggest that IBM would have been better off had antitrust been used to break it up. Leaving aside whether such a remedy would have been legal given the evidence at hand, it is probably not good policy for the government to decide which firms should be restructured to make them more responsive to future technological changes.

A more serious yet still incorrect rationale for the divestiture in *U.S. v. AT&T* is that vertical integration by monopolies is inherently bad, because of the propensity to "leverage" market power from the monopolized market into erstwhile competitive markets. Advances in the analysis of vertical integration have at least suggested some situations in which it may exacerbate market power. However, leveraging theories then and now still have to overcome the well-known argument that if a firm has a monopoly in one market, particularly one which provides an input on a "fixed proportions" (i.e., unit for unit) basis to downstream suppliers, it can already capture all of the monopoly profits that could be extracted after vertically integrating. If so, the only reasons to in-

---

2. *Motion of Bell Atlantic Corporation, Bell South Corporation, NYNEX Corporation, and Southwestern Bell Corporation to Vacate the Decree*, Appendix, U.S. v. Western Electric Co. Inc., and American Telephone and Telegraph Co., Civil Action 82-0192, U.S. Dist. Court (D. C.), July 6, 1994 (hereafter, "RBOC Affidavits"). The purpose of citing these affidavits is neither to criticize them nor to offer them as representative of expert consensus, but only use of them as an authoritative source of potential changed circumstances that would justify eliminating the MFJ.

tegrate vertically would be to realize cost reductions or other operating efficiencies that, while increasing profits, represent gains to the economy as a whole.

The justification for breaking up AT&T was, in fact, wholly consistent with the so-called "Chicago School" viewpoint that vertical integration and restraints are somewhere between benign and beyond the ability of government to determine which, if any, are malignant. William Baxter, the Reagan-appointed Assistant Attorney General who pledged to litigate *U.S. v. AT&T* "to the eyeballs" and engineered the divestiture, is not known for his populist or protectionist inclinations. The key factor that differentiated AT&T from IBM and other monopolists for whom multimarket activities have been the subject of antitrust scrutiny is not that it was big or potentially mismanaged. Possession of market power was a necessary condition, to be sure, but market power alone would not have been sufficient to justify requirements that AT&T divest its local operating companies, and that those divested companies by and large be prevented from manufacturing equipment or marketing information or long distance telephone services.

Rather, the crucial differentiating factor was that its monopolies in local telephone services were regulated, in particular, in such a way that rates for local telephone service were based on the imputed costs of providing those services. In different ways, the incentive for AT&T to continue as a multimarket firm, and to engage in the conduct for which it was accused, was to evade those regulatory constraints on the exercise of its market power in local telephone service. Since it could not directly extract monopoly profits through high prices to telephone subscribers and high interconnection charges to long distance companies, it allegedly looked to indirect means of extracting those profits through affiliated subsidiaries operating in markets for long distance service and telephone equipment.

The two tactics for regulatory evasion are a litany familiar to many telecommunications and antitrust lawyers, economists, and policy observers, even if there may be some confusion about the specifics.

Discrimination

The first of these tactics is known in the telecommunications community as "discrimination." In more traditional antitrust parlance, it might be better regarded as a Sherman §2 monopolization case, via a tie-in or control over an "essential facility." The story begins with the premise that AT&T's local telephone service was not only an important product on its own, but is a necessary adjunct or complement for the provision of a number of ancillary services. Among the most crucial examples of this were:

> A manufacturer of telephone switches and software, transmission lines, or sophisticated "customer premises equipment" (CPE), i.e., business and residential telephones and routing systems, may require information necessary to make one's products compatible with the local telephone service standards and technology.

Suppliers of "information services," i.e., data storage, processing and re-
trieval services including electronic mail and on-line data bases, need
timely and relatively noise-free connections to local telephone networks
in order to deliver their services to consumers. These providers might
also need technical specifications on how the local telephone networks
themselves switch, packetize, and transmit data, in order to optimize the
speed and precision of their information services.

Most notably, providers of "interexchange services," i.e., long distance
calling, generally have facilities that can route and transmit calls among
metropolitan areas. To actually connect residences or offices, the interexchange carriers (IXCs) have to connect to local telephone networks.

During the time at which AT&T's conduct was under scrutiny in the antitrust litigation, information services still existed more on the drawing board than in the
market. With regard to equipment, especially CPE, AT&T had attempted to use the
regulatory process first to bar entry outright, and then later to force its CPE competitors to install costly and generally non-functional "protective connecting arrangements."[3] AT&T's "protect the network" defense was hardly obvious and
sometimes implausible, particularly in its opposition to the marketing of a $10
plastic device one could place on a telephone handset, to make one's conversations
quieter and less intrusive to those nearby.[4] The most notorious examples of discrimination, however, involved denials, delays, and then degradation in the local
telephone network connections AT&T provided to MCI and other IXCs competing
with its long distance service.[5] More blatant examples of using the regulatory process to "raise rivals' costs"[6] probably could not be found.

Telecommunications policy researchers and practitioners have come to refer to
this family of cases as "discrimination," because AT&T typically treated its affili-

---

3. A good summary of this history is in BROCK, TELECOMMUNICATION POLICY FOR THE
INFORMATION AGE 79–101 (1994).

4. *Id.*, at 81.

5. United States v. AT&T, 524 F. Supp. 1336, 1348–52 (1981); Brock and Evans, *Predation: A Critique of the Government's Case in U.S. v. AT&T,* in BREAKING UP BELL: ESSAYS ON INDUSTRI-
AL ORGANIZATION AND REGULATION 41, 49 (David Evans, ed., 1982); Noll and Owen, *The
Anticompetitive Uses of Regulation: United States v. AT&T,* in THE ANTITRUST REVOLUTION
290, 304–06 (Lawrence White and John Kwoka, eds., 1992).

6. Noll & Owen, *supra* note 5, at 305. This phrase is taken from the now popular heading for a collection of theories covering practices by which a firm can increase its market power through strategies
designed to limit the ability of its competitors to compete. *See, e.g.,* Krattenmaker and Salop, *Anticompetitive Exclusion: Raising Rivals' Costs to Achieve Power Over Price,* 96 YALE L. J. 209 (1986).
Whether this theory is a productive addition to the antitrust policy maker's toolkit in unregulated contexts is a matter of some dispute. *See, e.g.,* Brennan, *Understanding "Raising Rivals' Costs",* 33 AN-
TITRUST BULL. 95 (1988); Coate and Kleit, *Exclusion, Collusion, or Confusion?: The
Underpinnings of Raising Rivals' Costs,* 16 RESEARCH L. ECON. 73 (1994).

ated equipment and long distance services more favorably than they treated unaffiliated competitors in those markets. In effect, however, the discriminatory episodes were tying arrangements, in which AT&T would essentially allow customers to use their monopoly local telephone exchange services only if they also used AT&T's CPE and long distance services. At first glance, these tie-ins conform to traditional "leveraging" Sherman §2 theories used to warrant concerns with tying arrangements, where market power in a monopoly was apparently used to monopolize complementary markets.[7] In *U.S. v. AT&T*, the tying monopoly was in local service, and the tied goods were CPE and interexchange service.

As such, however, a traditional tie-in theory of the case would have to overcome the objection that a tie-in is unnecessary to exercise market power that the monopoly already possesses over the tying good.[8] Depending upon one's perspective on antitrust policy, this objection is at least a nuisance, if not an insurmountable objection to proceeding with a case.[9] In *U.S. v. AT&T* and the related private cases, however, the plaintiffs could meet this general objection because state and Federal regulation of local telephone rates prevented AT&T from exercising its erstwhile monopoly power in local telephone markets. Without that regulation, AT&T would not have needed to monopolize interexchange service or CPE to maximize profits. It could have charged interexchange carriers monopoly access fees to originate and terminate calls, and it could have directly charged end-users high monthly dial-tone charges rather than maintain high equipment prices. Moreover, in principle it could charge even higher local service fees if complementary markets in CPE and interexchange became more competitive, with more efficient providers. Access and dial-tone regulation provided the economic incentive for AT&T to maintain its presence in those other markets, exercise its local exchange market power indirectly by discriminating against vertical competitors, and exploit that advantage by

---

7. POSNER, ANTITRUST LAW: AN ECONOMIC PERSPECTIVE 171–72 (1976).

8. *Id.,* at 173. Tie-in sales may allow a monopolist to extract more profit via that market power by enabling it price discriminate, but the effects on equity and efficiency from such price discrimination are difficult to predict in general. CARLTON AND PERLOFF, MODERN INDUSTRIAL ORGANIZATION 468–79 (1990). A recent game-theoretic rationale for tie-ins is that they may enable a firm to commit not to reduce output in the tied product market, because doing so would force it to forego profits in the tying product market. This commitment could deter entry in the tied product market. Whinston, *Tying, Foreclosure, and Exclusion,* 80 AMER. ECON. REV. 837 (1990).

9. BORK, THE ANTITRUST PARADOX 365–66 (1978).

charging higher prices for its affiliated services. The direct purpose of the tie-in is regulatory evasion; its manifestation is downstream monopolization.[10]

Cross-subsidization

The role of regulation is even more intimately connected with the second type of tactic motivating the divestiture and line-of-business restrictions: cross-subsidization. The gist of the cross-subsidization theory is creatively self-serving accounting.[11] Under traditional regulation, the rates for the regulated services such as local telephone subscription are, at least in theory, based on the regulated firm's costs for providing those services. If such a telephone company diversifies into other markets that use similar types of labor, resources (including financial capital), equipment, and technologies, such as long distance service, it could increase its profits by assigning the expense of those similar inputs to the regulated account. If its regulator treats those misallocated expenses as if they were spent on providing local telephone service, it will pass along those costs to the customers in the form of higher telephone rates. As far as the accounts are concerned, the firm will show earnings for telephone service equal only to the normal rate-of-return. It would get its profits from the unregulated enterprises, for which it had in effect charged some of the expenses to local telephone customers. While the profits show up in the unregulated markets and, thus, may appear to be the result of monopolization of those markets, they essentially are the direct result of exercising monopoly power over local telephony.[12]

As with discrimination, the incentives and consequences of cross-subsidization may still be widely misunderstood, because of a failure to recognize the importance of regulation. Just as discrimination could be viewed at first glance as an application of a traditional but problematic "leveraging" tie-in theory, so too could

---

10. The regulatory evasion theories depend on the premise that the other markets in complementary services are not tightly regulated. While the FCC set interstate long distance rates and state regulators set CPE charges, both of those were considerably above marginal cost. Consequently, AT&T would have the incentive to use its local exchange to inhibit competition in those markets, keeping those profit margins for itself. Brennan, Why Regulated Firms, *supra* note 1 at 773–75. Whether AT&T was justified in excluding competition, in order to preserve publicly mandated subsidies of local exchange service and deter inefficient entrants who could compete only because they would not have to pay those subsidies, was and continues to be controversial. Brock, *supra* note 3 at 143.

11. This distinguishes "cross-subsidization" as understood by telecommunications and antitrust analysts from the understanding in some economic contexts, where it is an indicator of misaligned pricing. The specific concern is whether prices are below "incremental" costs for some services and above "stand alone" costs for others. Faulhaber, *Cross-Subsidization: Pricing in Public Enterprises*, 65 AMER. ECON. REV. 966 (1975). In this antitrust context, the concerns are less sophisticated and, as we see below, need not necessarily imply that prices are "predatory" in some markets, at least in the Areeda-Turner sense of being below incremental costs.

12. The formal theory underlying this concern with cross-subsidization is in Brennan, *Cross-Subsidization and Cost Misallocation by Regulated Monopolists*, 2 J. REG. ECON 37 (1990).

cross-subsidization be described as an application of a traditional but problematic "deep pocket" predation theory, in which a wealthy telephone company uses its profits to support below cost pricing, to drive competitors out of its other markets. The standard argument against the "deep pocket" theory of predation is that the gross depth of the pocket does not affect the net profits of predation. If predation is profitable, a small firm should be able to do it. If it is unprofitable, as many would expect, the absolute size of the predator will not matter.[13] Furthermore, a firm will not raise its price its monopoly market just to fund predation in other markets; it presumably is already charging a price that generates maximum profits from its monopoly. If the firm is regulated so that the price in its monopoly market is based on reported costs, it then has an specific incentive to misallocate unregulated business costs to the regulated sector. Cost-based regulation may create a causal connection between output in the unregulated market and price in the regulated market that is generally absent in completely unregulated contexts.

For this reason, the prospect of cross-subsidization can make credible a threat to predate that would generally be not worth making in unregulated situations. However, cross-subsidization can be harmful without any adverse effect on price in the unregulated market. The regulated firm may be able to shift costs to the regulated side, raising rates there, without necessarily changing competitive conditions elsewhere. For example, suppose that a telephone company could cross-subsidize some of the up-front fixed costs associated with marketing of electronic data bases. If the cross-subsidy does not affect marginal costs, it will not lead the telephone company to increase its sales in the data base market. Whether the relationship between it and other data base providers is competitive or oligopolistic, overall market price and output would be unaffected by the cross-subsidy. Its only manifestation would be on the telephone company's data base profits, which would reflect, dollar-for-dollar, the results from raising rates in the regulated market.

Relief, Remedy, and Repeal

Regulators, legislators, and other policy makers were generally aware of at least the potential for competitive abuses following from discrimination and cross-subsidization. During the time the Justice Department was investigating and litigating against AT&T, these agencies offered to ease the concerns by imposing additional regulation of AT&T's internal operations to prevent these practices. The Antitrust Division, particularly after William Baxter became Assistant Attorney General,

---

13. As with almost any theoretical claim in industrial organization, one can find exceptions. TIROLE, THE THEORY OF INDUSTRIAL ORGANIZATION 379 (1989) argues that bankruptcy risk may make small firms less able to borrow. If so, large firms may have an incentive to cut price, keep the firms small, and prevent borrowing that would enable those small firms to compete in later periods.

sought the more draconian approach of forcing AT&T to divest its regulated local monopolies. After divestiture, the local monopolies would be stripped of the incentive and ability to evade regulation and inhibit competition in ancillary markets. A local telephone company has no reason to discriminate against unaffiliated long distance carriers or equipment manufacturers if it is itself not in those markets, since it would have no way to take advantage of that discrimination. Similarly, there would be no costs from unregulated operations to cover through local telephone rate increases if it could not enter unregulated markets. Realizing that its alternatives were intrusive internal regulation, continued restriction of its unregulated operations, and the likelihood that it would lose the case, AT&T found impossible to refuse the Justice Department's offer to end the case in exchange for divestiture.[14]

Despite whatever successes the divestiture has had in encouraging competition in equipment, long distance and information services markets, the residue that haunts us to the present day is that the principles warranting the break-up of AT&T into regulated monopoly and competitive components also warranted preventing the divested local telephone companies from re-entering AT&T's long distance and equipment markets, as well as the then nascent markets in information services. Accordingly, the MFJ imposed the notorious line-of-business restrictions on the RBOCs. Sensing their controversial nature, the court modified the Justice Department's proposal in a number of significant ways, allowing the RBOCs to sell CPE and directory advertising (Yellow Pages), requiring triennial reviews of the restrictions, and instituting a process by which the MFJ restrictions would be waived if the RBOCs could show that diversification would not harm competition in the markets they would be entering.

During the first triennial review in 1987, the Justice Department proposed lifting all of the line-of-business restrictions except that on interexchange service. The interexchange ban should be lifted if state regulators stripped the RBOCs of their exclusive franchises to provide telephone service. Among the many of the most important arguments in favor of this apparent reversal were that the burdens of MFJ enforcement, technological restructuring of telephone networks, and the high level of competition in the ancillary markets.[15] In addition, the Department, following the focus of the MFJ on competition in the entered markets, expressed the view that increase telephone rates caused by cross-subsidization were a regulatory problem, not an antitrust concern. In 1991, the court rejected the Department's arguments, citing a lack of competition in local telephony and a series of perceived abuses of their monopoly position. It retained the bans on manufacturing and interexchange

---

14. AT&T was also able to shed restrictions stemming from the original 1956 decree that kept it from providing anything other than communications services. U.S. v. American Telephone and Telegraph Company 552 F. Supp. 131, 226 (1982).

15. HUBER, THE GEODESIC NETWORK (1987). The "Huber Report" served as the factual foundation for the Department's reevaluation of the MFJ.

service, and reluctantly lifted the ban on information services only because neither AT&T nor the Department supported its continuation.[16] The present legal, regulatory, and legislative controversies are the legacy of that decision.

FROM WHERE WE STOOD TO WHERE WE STAND

The present compelling issue is the extent to which the rationale for the 1984 divestiture and the MFJ's line of business restrictions applies to the RBOCs in 1995. Reviewing that rationale, particularly the role that regulation plays, suggests four crucial questions:

> *Incentive:* Does the extent and nature of regulation continue to give local telephone companies an incentive to exploit their market power through diversification into other lines of business?

> *Ability:* Would local telephone companies have the market power necessary to be able to carry out discrimination and cross-subsidization tactics?

> *Monitoring:* Even if local telephone companies have this ability and incentive, do state and federal regulators, including antitrust authorities, have the means and motivation to detect and prevent anticompetitive conduct through diversified affiliates?

> *Efficiencies:* Are there benefits from allowing local telephone companies to diversify that outweigh the risks of harms from unpoliced discrimination or cross-subsidization?

Examining these questions in an absolute sense is difficult, if not intractable. A somewhat easier task, and one more relevant to practical legal and regulatory standards, is to see what circumstances may have changed since 1984 that would lead one to answer these questions differently in 1995.

Incentive

If the anticompetitive incentive to diversify is a byproduct of the regulation of local telephone service, changes in the incentive to diversify should depend on changes in the level or method of that regulation. One major change has been the widespread consideration and implementation of forms of "incentive regulation," such

---

16. AT&T, by this time a user rather than supplier of local telephone services, and as a potential victim of either RBOC discrimination and cross-subsidization or RBOC competition, opposed the Department's proposals for the elimination of the restrictions on manufacturing and interexchange services.

as price-caps. Despite the many names and nuances, the defining idea of incentive regulation schemes that differentiates them from traditional rate-of-return regulation, is that the price the regulated firm can charge for its services is independent of the costs over which it has discretion. Under traditional regulation, rates are (in theory) set in such a way to generate just enough revenue to cover the costs of providing the service. When rate-of-return regulation works according to design, a firm has no incentive to control expenses in the short run or to reduce them in the long run through innovation. Any reduction in cost would bring about a reduction in price, stripping the firm of any profit it would receive from cutting cost.

The "incentive" in incentive regulation comes from the inference that if a regulated firm cannot raise rates if costs go up, and if it will not have to cut rates if its costs fall, then it will keep at the margin any profits it makes if it cuts costs. Consumers share in the expected productivity gains from *ex ante* imposition of productivity-related rate decreases, but these decreases are, in principle, divorced from actual reductions in costs, to preserve the incentive to economize.[17] Moves away from cost-of-service regulation to direct price controls or incentive regulation would reduce or eliminate incentives to cross-subsidize unregulated services, by taking away ratepayer funding of unregulated operations.[18] Regulated rates would not be raised to cover misallocated costs of those services. Over thirty states are considering or implementing some forms of incentive regulation of local telephone companies.[19]

The effectiveness of price-caps requires a public commitment to divorce regulated rates from prices, which is difficult and unlikely for the government to achieve.[20] If the regulated firms appear to be making a great deal of money because of the cost-cutting incentives created by price caps, there will be considerable political pressure to force a reduction in prices and, perhaps, a return to cost-

---

17. Kwoka, *Productivity and Price Caps in Telecommunications*, in PRICE CAPS AND INCENTIVE REGULATION IN TELECOMMUNICATIONS 77 (Michael Einhorn, ed., 1991). There is some evidence from intrastate long distance markets that incentive regulation leads to lower prices. *See* Mathios and Rogers, *The Impact of Alternative Forms of State Regulation of AT&T on Direct Dial Long Distance Telephone Rates,* 20 RAND J. ECON. 437 (1989). A question, difficult to answer through econometric testing, is whether an observed relationship between lower rates and incentive regulation reflects the effect of the incentive regulation on costs and prices, or whether state regulatory expectations on the intensity of competition lead it to adopt less rigorous forms of regulation. If the latter is the case, imposing incentive regulation where regulators have chosen not to do so might not lead to reduced prices, despite the observed correlation between the two.

18. Brennan, *supra* note 12. RBOC affidavits by Arrow and Carlton at ¶20, Crew and Kleindorfer at ¶13, Milgrom and Roberts at ¶49, Perloff and Karp at ¶125–26, and Sappington at ¶30.

19. RBOC Affidavits by Arrow and Carlton at Appendix 2.1–2.2, Mitchell and Vogelsang at ¶70, Wilk at Appendix 2.

20. *See* Brennan, *Regulating by 'Capping' Prices,* 1 J. REG. ECON. 133, 142 (1989); Baron, *Information, Incentives, and Commitment in Regulatory Mechanisms: Regulatory Innovation in Telecommunications,* in PRICE CAPS AND INCENTIVE REGULATION IN TELECOMMUNICATIONS 47 (Michael Einhorn, ed., 1991).

based regulation. The imposition of federal regulation of cable rates in 1992, as a response to the price and profit increases that followed the 1984 Congressional preemption of price controls built into local cable franchise contracts, is a useful example of this tendency. In the other direction, a regulated firm that loses money, when costs fail to fall as much as was anticipated or allowed for in advance, will be expected to exert political pressure to raise its rates.

Not only might we expect political institutions to be responsive,[21] but legal standards guaranteeing the opportunity to earn a "just and reasonable" return on investments[22] reflect judicial dispositions to protect the firms as well. Such dispositions appropriately reflect the need to make the government honor its commitments to regulated firms that, having installed expensive plants and equipment, are vulnerable to post-construction regulatory decisions to set prices so low that only operating costs, not those capital investments, are covered. In any event, if legal constraint or political pressure continues to tie costs to rates, moves to incentive regulation may be little more than rate-of-return regulation with a time lag between the change in costs and the change in rates. While lengthening the lag between cost and rate changes provides less incentive to cross-subsidize, it does not eliminate that incentive altogether.

It is also important to note that while price caps or incentive regulation can eliminate the incentive to cross-subsidize, it does nothing about the incentive to discriminate against unaffiliated competitors in unregulated markets. The latter incentive comes about because effective discrimination would tie purchase of the unregulated service to purchase of the regulated service. This tactic would allow the firm to add the otherwise prohibited monopoly premium for the regulated service to the price of the unregulated service. Discrimination prevents competitors in the unregulated market from unraveling the tie by giving consumers the opportunity to avoid it. Incentive regulation combats cross-subsidization only because that tactic depends not only on the fact of regulation but on its form, particularly, the policy that changes in reported costs affect changes in rates.[23]

The inability of price caps to eliminate the incentive to discriminate, and the difficulty of making commitments, suggests that the real way to eliminate the need to restrict lines-of-business to prevent evasion of regulation is simple: eliminate the regulation. While there is some discussion of deregulating the local exchange, such moves appear

---

21. Peltzman, *Toward a More General Theory of Regulation*, 19 J. LAW ECON. 211 (1976).

22. Federal Power Commission v. Hope Natural Gas, 320 U.S. 591 (1944).

23. A variant on the cost misallocation story is that a regulated firm could vertically integrate upstream, and sell inputs to itself at inflated prices, causing regulated rates to rise, with the attendant profits realized on the books of the upstream subsidiary. Since this tactic also requires cost-based regulation to be effective, incentive regulation would combat it as well. *See* RBOC Affidavit by Fisher, ¶25.

not to be widespread.[24] This is unfortunate, not only because it fails to remove the clear cause of the incentive for anticompetitive diversification, but because it implies that local exchange companies may retain the market power necessary for giving them the ability to discriminate or cross-subsidize. To that question we now turn.

Ability

Regulation would not matter if local telephone companies faced competition. If long distance companies could connect to any number of local exchange companies to originate and complete calls, the regulated incumbent could not create a market advantage for its affiliated interexchange carrier by delay or degrading interconnections. The interexchange competitors would look elsewhere for access. If consumers could choose among local exchange competitors for telephone service, a regulated incumbent could not pass along rate increases constructed with the help of misallocated costs. Consumers would evade the cross-subsidization by turning to other carriers who did not try to raise prices, preventing not only higher local rates but eliminating the threat that cross-subsidization would be used to fund predation.

An important and yet to be fully appreciated development since the 1984 divestiture is that technological change and unbundling of local telephone services has apparently created separate service markets within the "local exchange" rubric, with potentially different competitive and regulatory circumstances.[25] The "old" theory rationalizing the divestiture and the line-of-business restrictions by and large treated local exchange service as a monolith. In fact, the component parts of that local exchange service all may constitute separate markets, with separate competitive conditions. Among the potential separate or separable markets, with potential competitors listed, are the following:

> *"Local loops,"* i.e., connections from customer premises to a switch that allows calls to be routed throughout the local exchange area, including to points of connection with long distance carriers. The telephone company currently provides these connections primarily with copper wires. Cable television cable, cellular radio-telephones, and planned wireless "personal communications systems" are actual or potential competitors in some degree.

> *Bulk transport*, i.e., unswitched private-line transmission capacity between two points. Telephone companies provide this service either on a contractual special construction basis or by committing capacity of their

---

24. The RBOC Affidavit by Perloff and Karl raises the issue, but mentions only discussion of the issue in California (¶126-28). Appendices 2.1 and 2.2 of the RBOC Affidavit by Arrow and Carlton indicates that only Nebraska has adopted a general deregulatory policy; Indiana and South Dakota have deregulated small telephone companies, with less than 6,000 or 10,000 lines, respectively.

25. *See, e.g.,* RBOC Affidavit by Mitchell and Vogelsang at ¶3.

local networks to ensure that these connections are not congested with other telephone traffic. The most notable examples of such a service involve direct connections between high volume users, e.g., large companies or office buildings, and interexchange carriers. Fiber-optic technology has opened up this bulk transport market to so-called "competitive access providers" (CAPs[26]), saving their customers or the need to pay local telephone company charges for this service.[27]

Local switching networks that collect and route traffic throughout the local exchange. A limited amount of switching capacity has faced competition for some time, in that multiline business users can substitute an on-site switch for the local network switch for calls within the firm's premises. In a somewhat similar manner, a cellular telephone company's switches can route calls among its subscribers without going through the local exchange. The fiber-optic lines CAPs provide to central business districts could be used for routing calls in those areas.[28] Last, and perhaps most important, these technologies may render cable networks convertible into switched networks, despite the fact that cable networks are typically designed to send signals in one direction from a single central point.

There may be other separate markets relating to billing, operator services, and other enhancements to local service, such as switch-based voice mail or call-forwarding.

Not all of these markets appear to be equally competitive. Bulk transport appears to be the most competitive; loops and networks less so. Moreover, the ability to enter and compete in some or all of these markets depends on thorny interconnection price and policy issues that state and federal regulators have only begun to confront, much less resolve. However, the possibility that local exchanges service combines separate markets means that the role of regulation and the extent of market power need to be considered on a market-by-market basis. In particular, evidence of "market power" need not imply that the line-of-business restrictions are justified, nor does evidence of "competition" imply that the restrictions are unjustifiable.

Consider the interexchange service question, and assume that bulk transport is competitive while local loop and switching markets remain regulated monopolies. Suppose first that control over bulk transport is necessary to carry out discrimination, or that the only interexchange costs that could be falsely represented are

---

26. Not to be confused with "price caps."

27. RBOC Affidavits by Hausman at ¶14, Mitchell and Vogelsang at ¶12; Brock *supra* note 3 at 247–54.

28. An interesting related development is that the capacity of fiber-optic lines may be so great that one could design networks so all calls go through all lines to all locations, where on-site signal identifiers pull out only those calls intended for that destination. Such a network would provide the equivalent of switched service without the switches.

those similar to costs incurred in the provision of bulk transport. If so, continued regulated monopoly over local loops becomes irrelevant, since the RBOC would no longer have the relevant market power necessary to discriminate or cross-subsidize. However, suppose that costs could be shifted onto the loop or switching sectors, or that loops and switched service provision might be denied or delayed to interexchange carriers or their customers. In this case, competition in the bulk transport market is irrelevant; the ability to discriminate or cross-subsidize persists. In general, competition in some markets may not warrant lifting of restrictions if the crucial markets will remain regulated monopolies for the foreseeable future. Unlike 1984, when local exchange service appeared to be a permanently regulated impregnable monopoly, resolution of this issue now requires a much more careful analysis of exactly how a local telephone company could discriminate and cross-subsidize, and whether those tactics are susceptible to regulatory monitoring and prevention.

Monitoring

Continued regulation-created incentives to vertically integrate and discriminate and cross-subsidize, and continued power in relevant regulated local exchange markets that creates the ability to disadvantage competitors and to pass through costs to consumers, do not warrant antitrust-related concern, if the regulators are able to monitor and police RBOCs to prevent such conduct. The underlying premise of the divestiture was that regulators were unwilling or unable to deter discrimination or control cross-subsidization during the decades when AT&T's allegedly was acting anticompetitively. The question, therefore, is whether the situation has changed. This involves two separate inquiries as to whether that ability or willingness has changed.

Since cross-subsidization is essentially an exercise in misallocating cost, improved ability to counteract it would involve better accounting techniques. Since the divestiture, the FCC and some of the states have imposed more detailed cost reporting rules.[29] The courts have cast doubt on whether these improvements are a material change, vacating and remanding the FCC's 1987 Computer III rules, in part because the FCC did not explain why it had changed its conclusion in its 1980 Computer II proceeding that accounting controls were ineffective.[30] Moreover, accounting controls will not work if networks are designed in such a way to maximize the costs that are shared and fundamentally impossible to attribute specifically to either the regulated or unregulated service. If these shared costs are charged to the regulated side, we can get cross-subsidization, despite accounting

---

29. Brock, *supra* note 3 at 227. *See also* RBOC Affidavits by Farmer at ¶¶8–11, 25–31; Rivera, Firestone, and Halprin at ¶66, 72, 80; Sappington at ¶37; Wilk at ¶18.

30. California v. FCC, 905 F.2d 1217 (Ninth Circuit, 1990).

"safeguards" ensuring that no post-design costs attributable to the unregulated sector are misallocated to the regulated side of the business.[31]

There maybe more cause for optimism regarding discrimination. The FCC's Computer III proceeding also initiated a set of rulemakings leading to prescriptions that local telephone companies provide unbundled "open network architecture" and "comparably efficient interconnection" arrangements. These rules would ostensibly allow unaffiliated competitors in local exchange markets or downstream telecommunications services to get the same treatment the telephone companies give their own affiliates, without forcing competitors to purchase service they do not need.[32] These reforms have faced some skepticism.[33] Moreover, rules that mandate equal treatment leave an telephone free to "discriminate" to some degree against firms with technologies and services that would work better with a different type of connection than a phone company gives itself.[34]

A second reason to be more hopeful with discrimination is grounded in the belief that regulators have been effective at monitoring discrimination since the divestiture. The two prominent contexts in which regulators have shown their ability are the "equal access" requirements between interexchange carriers and the RBOCs, and the interconnection of cellular carriers to local networks.[35] In neither case, however, is the analogy perfect. Equal access for interexchange carriers may have been effective and easy to enforce just because the MFJ's restrictions eliminated any incentive the RBOCs might have had to discriminate.[36] Whether enforcement would be equally effective if the RBOCs had an incentive to favor an

---

31. Brennan, *supra* note 12, at 42–44; Baseman, *Open Entry and Cross-Subsidization in Regulated Markets,* in STUDIES IN PUBLIC REGULATION (Gary Fromm, ed.) 329 (1981).

32. Brock, *supra* note 3 at 224–26. *See also* RBOC affidavits by Sappington at ¶15-16.

33. COLE (ed.), AFTER THE BREAK-UP: ASSESSING THE NEW POST AT&T DIVESTITURE ERA 56 (commentary by William McGowan), 165 (commentary by Nina Cornell), 345 (commentary by Dale Hatfield) (1991).

34. Brennan, *Divestiture Policy Considerations in an Information Services World,* 13 TELECOM. POLICY 243, 251 (1989). Ironically, a potential virtue of these unbundling rules is that they may force RBOCs to adopt standardized technologies, that would prevent designing systems in such a way to create artificially shared costs of providing regulated and unregulated services that might be shifted to the regulated sector. *Id.,* at 252.

35. RBOC Affidavits by Arrow and Carlton at ¶29; Hausman at ¶21; Rivera, Firestone, and Halprin at ¶30, 44.

36. In theory, equal access regulation should be unnecessary. The two practical reasons for requiring it are first that as long as local telephone service is regulated to just cover costs, local telephone companies have no affirmative profit motive to ensure that everyone is efficiently connected. More specifically, there was concern at the time of the divestiture that the RBOCs would honor old family ties and continue to favor AT&T despite the divestiture. This latter concern appears to have proven quite unwarranted.

affiliate is another question.[37] The cellular experience speaks more readily to that issue, but the fact that regulators structured cellular with only two suppliers per market, with limited spectral capacity, raises some doubts as to how competitive the industry would really be. If the gains from duopoly are sufficiently great, there may have been little reason for local telephone companies to resist antidiscrimination requirements, especially if engaging in discrimination in this market, and getting caught, might lead the government to extend restrictions on their entry into other markets.

Recent efforts to formulate and enforce antidiscrimination and proper cost allocations rules suggest that whether regulators are more inclined to monitor may be more important than whether regulators are better able to monitor. The *U.S. v. AT&T* history is instructive. Cross-subsidization was never detected as such. Instead, AT&T was accused of "pricing without regard to cost," i.e., acting as if it could cover any losses from undercutting competitors by shifting costs to its regulated monopolies. The alleged discrimination against competing CPE and long distance providers was open and notorious; the main dispute was about whether the conduct was excused by FCC regulation, or a manipulation of it. In either case, a key factor was that the regulators were not particularly sensitive to the need for vigilance to protect competition in ancillary markets. Now that such competition is active rather than merely anticipated, and now that anticompetitive abuses are more widely understood, the threat of regulation and enforcement of unbundling rules may be a more credible and effective deterrent to monopolistic conduct.[38]

If attentive regulators can thwart overt discrimination and apparent predation, then anticompetitive conduct depends on nearly paradoxical circumstances.[39] For discrimination to be effective, consumers must recognize its effect, otherwise they will not pay a premium for service from the favored affiliate, and the unaffiliated providers will not be put at a competitive disadvantage. However, for discrimination to be successful, the regulators must not be able to recognize it. A similar "between a rock and a hard place" argument applies to cross-subsidization. Regula-

---

37. A pricing issue related to discrimination is whether the RBOC would give interconnection access discounts to its affiliated long distance carrier. Even if the RBOC is nominally required to charge its affiliated interexchange carrier the same access fee it charges its unaffiliated competitors, it may act as if it is charging itself only the marginal cost of providing access, unless its interexchange carrier is acting as an independent profit center. Without that independent accountability, a telephone company may have no reason to regard a fee it charges to its affiliated interexchange carrier as a cost.

38. A related argument is that a benefit of the divestiture was the creation of seven local telephone companies. Regulators might be able to regulate better if they can compare the performance of the firm they are regulating to those that they are not. Whether such "benchmarking" causes movement toward more efficient performance or regression to an inefficient mean is difficult to predict. The similarities across the RBOCs in their incentives and abilities as described above may outweigh differences among them that their regulators can exploit.

39. The following arguments if from Faulhaber, TELECOMMUNICATIONS IN TRANSITION 118–120 (1987); *see also* Brennan, *supra* note 34, at 247–49

tors must be sufficiently diligent to pass costs on to ratepayers, but insufficiently diligent or able to ascertain whether those costs were actually employed in the provision of the regulated service.[40] It may be optimistic to conclude that the regulators can monitor discrimination and cross-subsidization to eliminate any risk of monopolistic abuses, but the nearly open invitation to engage in such abuses facing AT&T in from the early 1950s to the late 1970s is no longer on the table.

Efficiencies

The line-of-business restrictions impose two possible categories of economic losses in the telecommunications industry and the economy at large. The first is that they may prevent the RBOCs from realizing economies of scope between local exchange services and equipment manufacturing or interexchange service. The second, and more interesting case, is that keeping the RBOCs out of ancillary monopoly or oligopoly markets may make those markets less competitive than they otherwise would be.[41]

While the *U.S. v. AT&T* litigation never officially reached the "relief" stage, when the pros and cons of the divestiture would have been contested, economies of scope issues were paramount at the time. These included contentions regarding the efficiencies from complete network management, "one-stop" shopping, common switches and lines for both local and long distance service, and complementarities in research and development. One of the mottoes of AT&T during the litigation was "If it ain't broke, don't fix it." However, as William Baxter put it in an interview five years after the divestiture, "The decree [MFJ] implicitly made a wager that the regulatory distortions of those portions of the economy, which could have been workably competitive, yielded social losses in excess of the magnitude of economies of scope that would be sacrificed by this approach."[42] Current claims of scope economies by and large make arguments that would have applied to the "sacrifice" side of Baxter's balance at the time of the divestiture. They may suggest that breaking up AT&T was a mistake in the first place, but they offer little guidance to courts or policy makers having to ascertain whether 1995 is sufficiently different from 1984 to warrant a change in policy.

A much more interesting possible changed circumstance involves whether the markets from which the RBOCs were banned have become as "workably com-

---

40. Some commentators point out that political pressures would also discourage regulators from shifting costs in such a way as to raise local telephone rates. *See, e.g.,* RBOC Affidavits by Wilk at ¶ 17. Of course, this was true prior to the divestiture, and it did not seem to deter "pricing without regard to cost."

41. The identification of these losses and estimations of their magnitude relative to the gains from preventing cross-subsidization are examined in Brennan & Palmer, *Comparing the Costs and Benefits of Diversification by Regulated Firms,* 6 J. REG. ECON. 115 (1994).

42. Cole, *supra* note 33, at 30.

petitive" as Mr. Baxter and the advocates of divestiture anticipated. The greatest controversy has been whether the long distance market is competitive, dominated by AT&T's price leadership,[43] or a tacitly colluding cartel.[44] The key piece of evidence seems to be that prices in the long distance market, net of interconnection payments to local telephone companies, seem not to have fallen as fast since the divestiture as they did prior to the divestiture, if at all.[45]

As with almost any other empirical question in economics, ferreting out the signal from the noise is not easy. Since AT&T was, at least nominally, a rate-of-return regulated firm in long distance as well as local markets, with a depreciating rate base on which to reap earnings and with growing demand, one would expect that its long distance rates at the time of the break-up might have been near the competitive level and declining. Moreover, the breadth of service offerings and discount packages interexchange carriers make available to business and residential customers implies that identifying a "price" of long distance service to compare with interconnection fees may be tricky, to say the least. Any benefits of RBOC entry into proscribed markets depends crucially on whether they will be net additions to the market, rather than as replacements for either current competitors or for other firms who would have entered in their stead.[46] Finally, any benefits of RBOC entry may depend on not just on its presence, but whether that entry reduces AT&T's role as a price leader or disrupts any tacit collusion. Such benefits cannot be guaranteed.

THE EPHEMERAL BOTTOM LINE

Any prescriptive theory in economics is only as good as its assumptions are valid. Consequently, confidence in prescribing that local telephone companies remain out of ancillary markets would have to be weaker now than in 1984, but still not compellingly so. Regulatory reform may reduce the incentive to cross-subsidize, but failure to deregulate still perpetuates the incentive to discriminate. The disaggregation of local exchange service into separate product markets, with competition primarily in bulk transport and connection to interexchange carriers, offers some hope that the ability to exercise market power via discrimination or cross-subsidization may have fallen in the last decade. However, competition in the full panoply of local services is certainly not yet strong enough to have persuaded state and federal governments to deregulate local service. Experience in monitoring telephone company conduct and efforts to devise new accounting controls may be

---

43. RBOC Affidavit by Hausman at ¶ 7, 34; Schmalensee at ¶ 4, 12; Taylor at ¶ 8.
44. RBOC Affidavit by MacAvoy at ¶ 39.
45. RBOC Affidavit by Schmalensee at ¶ 10, table 1; Taylor at ¶ 18.
46. Brennan & Palmer, *supra* note 41, at 131–34; even if there is net entry, diversification is likely to reduce net economic benefits unless there were three or fewer firms in the entered market.

less relevant or significant than the perception that the regulatory atmosphere is far less conducive to anticompetitive consequences than it was prior to *U.S. v. AT&T*. Any remaining risk of anticompetitive conduct might be worth taking if it would increase competition in other markets. The evidence of competition in interexchange service is apparently not universally persuasive, but it is difficult to tell whether that market is sufficiently oligopolistic to justify vacating the MFJ.

A final consideration may suffice to sway many of us sitting on the fence. If telecommunications industry participants believe that the MFJ's days are probably numbered, that in and of itself might be a good reason to drop it. Policies, such as the line-of-business restrictions, that affect long-run business and investment plans are effective only to the degree that the government can credibly commit to them. Public commitment to particular policies, regulations and laws is not impossible; our monetary and commercial systems rely on it. In this instance, however, it is hard to argue that the government is guaranteeing that the restrictions in the MFJ will be maintained. Doubts began when Judge Greene allowed the RBOCs to provide CPE and introduced the triennial review and waiver process, and were greatly exacerbated by the Justice Department's support of repealing most of those restrictions by 1987. Since then, there has been perpetual discussion of regulatory or legislative efforts to take the MFJ out of the hands of the antitrust courts.

If credible commitment to continue the line-of-business restrictions is absent, the present state of affairs may be the worst of all possible worlds. Continuing to implement the MFJ restrictions deters entry by the RBOCs, while the expectation of future revocation may deter entry by RBOC competitors, out of fear that the RBOCs are either more efficient competitors or prepared to reenact in a more subtle manner the anticompetitive scenarios that led to the divestiture in the first place. Consequently, the current compromised environment could well be the least competitive of all. The ideal solution might be for the government to reestablish a credible commitment to maintain the MFJ. If that is impossible, and certainly if it is unwise, the best solution may be to go to the other extreme and vacate the MFJ altogether.

Whether the validity of the theory itself or the credibility of the government's commitment should be the factor determining the MFJ's survival, the legal, political, and economic resolution of the controversy will determine on an analysis of that theory's assumptions. Identifying the key issues—incentive, ability, monitoring effectiveness, and countervailing efficiencies—and the facts that might determine how they have changed since the 1984 divestiture should help lead to better telecommunications policy making as we near the 21st century.

ACKNOWLEDGMENTS

This chapter was originally published in the Fall 1995 issue of the *Antitrust Bulletin*; permission to reproduce it here is gratefully acknowledged. This paper was written while the author was a Gilbert White Fellow at Resources for the Future; RFF's support is gratefully appreciated. The author also has consulted with the Antitrust Division of the Department of Justice on telecommunications policy matters. Nothing here represents the position of RFF, the Department of Justice, or any of their officials or personnel. The author appreciates very much the comments of Robert Thorpe, Molly Macauley, David Loomis, Arthur McGrath, and participants in the 14th Annual Rutgers Advanced Workshop in Regulation and Public Utility Economics. All opinions and errors remain the author's responsibility.

The Rochester Local Exchange Market Nine Months Later

Richard Stannard
New York State Public Service Commission

In 1993, Rochester Telephone Corp. was looking to form a holding company to give itself more flexibility in the competitive telecommunications markets, as well as in the financial markets. Roughly at the same time, the New York State Public Service Commission was beginning to consider how best to encourage competition in the local exchange markets in New York. In addition, there was a growing recognition on our part that the telecommunication industry was generally entering a period of declining costs, and that as surrogates for competition it was our responsibility to ensure that consumers received a fair share of the productivity gains that companies were beginning to achieve. Thus, the timing was ripe for the convergence of these three priorities into an integrated framework.

The Rochester market was also uniquely attractive as a test bed, given its compact operating territory and good cross-section of customers, with urban and rural communities and small to large business customers.

Given this backdrop, the Open Market Plan (OMP) is the result of an innovative plan filed by Rochester Telephone Corporation, which was ultimately modified through a process of negotiations involving approximately 20 parties.

MAJOR ASPECTS OF THE OPEN MARKET PLAN

The major aspects of the OMP are: (a) the establishment of a holding company structure, (b) implementation of performance-based incentive mechanisms, and (c) a framework for opening the Rochester market to competitive entry for local exchange services.

Our concerns with the holding company structure included potentially increased utility costs due to affiliate transactions, a diversion of capital to the detriment of utility customers in service quality and availability, and a diversion of managerial attention. All of these were satisfied by the numerous protections built into the plan.

The performance-based incentive mechanisms include a tight service quality plan with monetary penalties for failure, no provision for recovery of inflationary or exogenous cost increases, and no earnings cap. In addition, the company is required to reduce revenues by $21 million over the term of the plan, or by 7% (based on 1994 revenues).

The competitive enhancements, which are discussed in greater detail later, provide for development of facilities-based competition and establish a wholesale–retail rate structure through which all of Rochester's services, including bundled residential exchange service, may be resold.

Although the OMP was approved as an integrated framework, each proposition (i.e., the holding company, the performance mechanisms, and the measures to open the local exchange market) was designed to operate on a more or less stand-alone basis. In other words, the success of the performance aspects of the plan do not depend solely on the level or growth of competition. Similarly, Rochester's local service customers and competitors are shielded from any potential harmful fallouts from a holding company structure by the price and service protections built into the overall plan, as well as by competition.

In opening the local exchange market to competition, the focus was on encouraging the development of competing networks. The significant issues that emerged from this focus were: the measures required for the physical interconnection of multiple networks, compensation arrangements for the exchange of local traffic, interim and long-term number portability schemata and availability of bottleneck services and databases at reasonable terms, conditions, and prices (e.g., the ability for competitors' customers to be listed in Rochester's directories and to be found in directory assistance and 911 databases). From our vantage point, penetration by any type of competitor into the Rochester local exchange market seemed difficult. Rochester's rates are comparatively low, at least in New York, and Rochester has a tradition of providing excellent service.

We did not wish to handicap the race. The cable company was starting with its networks fully deployed, or nearly so, while others needed to start with full resale of the encumbent's services, as AT&T is now doing in the Rochester market. Our intent was that each competitor be given a reasonable opportunity to enter and survive in the local telecommunications markets.

We now understand more clearly that full resale is, for some competitors, a crucial first step into the market on the way to developing a full-fledged competing network. To gain a market presence, an entity may begin by reselling virtually all of the incumbent's services, as AT&T is now doing in Rochester. This activity may be followed by the placement of a central office switch and the need to resell only certain components of the incumbent's network (e.g., links that may be used to connect the consumer to the competitor's own switch). Thus, the terms, conditions, and prices associated with resale have become elevated in importance.

ROCHESTER MARKET NINE MONTHS LATER

At the time of this writing, it has been approximately 9 months since the OMP was implemented. We can make the following observations about the market at this time.

There are 4 resellers operating in the market: AT&T Communications of New York, Inc. (AT&T), Citizens Telecommunications Company of New York, Inc., ICS Telecom, Inc., and Frontier Communications of Rochester, Inc. (a company wholly owned by Rochester's holding company, Frontier, which is certified as both a facilities-based provider as well as a reseller). Together, these resellers have attracted about 5% of Rochester's embedded exchange access line customer base. The 5% does not include the approximate 56,000 Centrex lines that were transferred from Rochester to Frontier pursuant to the terms of the plan. Excluding Frontier, the access line market penetration for non-Rochester-affiliated resellers is about 2.9%. AT&T was marketing solely to residential customers. AT&T has indicated that it has suspended marketing activities (this is discussed further later). Time Warner is the only facilities-based competitive carrier active in the market at this time, but on a limited basis. The company is being very cautious and measured with its approach to the market.

We understand that MFS intends to enter the market shortly as a full-service, facilities-based carrier, despite its criticism of certain aspects of the plan.

With respect to Rochester Telephone Corporation, 1995 operations yielded the following: On January 1, 1995, the company implemented an $11 million revenue reduction through the elimination of residence touch-tone charges, and reductions to business local usage charges and carrier access charges. The company's net income through July 1995 was up by $6.2 million, or by 22% over the same period in 1994.

CONSUMER IMPACTS

There are a number of ways to assess the impact of the OMP on consumers, but price and service are, perhaps, the most significant. Currently, Rochester, Frontier, AT&T, and, to a more limited extent, Time Warner are serving the residential market. The monthly retail price for flat rate, individual service, including touch-tone service in the Rochester metropolitan area is $12.96 (exclusive of the FCC subscriber line charge) for Rochester, Frontier, and AT&T. Time Warner's price is $11.60 (exclusive of the FCC subscriber line charge) per month, and its prices for more discretionary services are typically 10% below Rochester's. Time Warner conducted a trial in approximately 50 residential apartments during the first part of 1995. At the conclusion of the trial, the vast majority of customers elected to stay with Time Warner rather than returning to Rochester. Presumably, these customers find the combination of Time Warner's price and service to their satisfaction. There is also some indication that customers who subscribe to both Time Warner's telephone service and cable-TV services like having a single point of contact for customer service issues and billing.

With respect to business services, the picture is similar. AT&T does not serve the business market, but Citizens does. Rochester's and Frontier's retail monthly price for a measured rate business line is $11.72; Time Warner's is $10.70, and Citizens' is $20.00. Time Warner also offers a flat rate business service for $40 per month, and Citizens' monthly flat rate option is $45.

Rochester's quality of service is becoming an issue with its customers. Each month, the company conducts customer satisfaction surveys with a sample of customers who have had recent transactions with the company (e.g., changes in service, billing inquiries, etc.). In addition, the company does an annual survey using a sample from the general body of customers. Both show a material decline in customer satisfaction with Rochester Telephone Corporation since the outset of the OMP. These results are borne out by an increase in consumer complaints to the commission concerning various aspects of the company's performance. Complaint levels for the first 6 months of 1995 were up 33% over levels for the same period in 1994. The complaints are primarily about a new billing format and the inability of customers to reach the company. There is no indication that the implementation of the OMP caused any significant confusion in the market or service problems, although it may have generated an increase in the number of calls to the company, for which it is having a problem in handling.

Rochester's service quality declined in various categories during 1995. Although we have repeatedly brought to the company's attention poor answer time performance levels for business office and repair service bureaus, the company has been unable to provide objective-level service on these measures. Under the service quality component of the OMP, the company has to maintain a yearly minimum objective service point requirement of 85% for four major service quality measurement categories. These are customer trouble report rate, percent out-of-service over 24 hours, percent missed repair appointments, and an aggregate of various answer time and installation performances. As of September 1, 1995, Rochester's performance on these measures was only at 80.7%. For comparison purposes, we looked at Rochester's performance for the first eight months of 1994; the 1994 performance was at 82.4%. Unless significant service quality improvement occurred during the remainder of 1995, the company was likely to fall short of its commitment of maintaining the 85% service quality objective level; and it may have become liable for some penalties.

MAJOR ISSUES CONCERNING THE OPERATION OF THE PLAN

In the previous 9 months, several issues have arisen concerning the operation of the plan. Some of these issues were anticipated, and others were not.

WHOLESALE DISCOUNT

Among resellers, the issue of the wholesale discount is, without a doubt, the most contentious. All, save Rochester's affiliate, Frontier, complain that the existing 5% differential between Rochester's wholesale rates and its retail rates is far too thin, and that they cannot make any profit or compete effectively against Rochester's retail rates. AT&T claims that it has suspended its marketing activities in Rochester because the current rate structure makes providing service a losing proposition. We know that Rochester is negotiating changes in the discount structure with the resellers—in particular with AT&T. The company wants a business solution to this issue, rather than an imposed regulatory solution. Thus, Rochester's offer will have to "satisfy" the resellers that they would not do better in a battle before the PSC because the carriers probably will never admit they are fully satisfied.

As previously mentioned, our focus was facilities-based competition. We continue to feel that this form of competition is the most powerful. However, had we understood the importance of resale as at least a startup tool, a great deal more attention would have been paid initially to wholesale–retail issues—in particular to the wholesale rate structure. In retrospect, the 5% discount looks thin and, perhaps, it should have been disaggregated (i.e., the uniform 5% applies to residential exchange access lines, which are priced below embedded cost, as well as to highly profitable discretionary services and complex competitive business services, where the marketing and sales costs are considerable).

When considering the merits of the plan, there was little evidence in the record on which to construct a more thoughtful wholesale rate structure. Some parties, like LDDS, pointed out perceived weaknesses, which Rochester disagreed with. Consequently, the commission directed us to report back to it any changes that might be required after some actual experience was gained.

As a result, we are about to initiate a process to review pertinent facts, and to hear from the parties concerning their views on the wholesale rate structure for the purpose of revising the structure by the end of the year. The commission has also initiated a proceeding designed at implementing a wholesale rate structure for New York Telephone and the rest of the local exchange carriers in the state so that their services may also be resold at commercially viable prices.

SERVICE ORDER PROCESSING
AND INFORMATION FLOW BETWEEN CARRIERS

The OMP included a provision that would allow carriers taking service from Rochester electronic access to Rochester's customer records and its order entry and repair record databases. The databases were to be partitioned so that each carrier had access to only its customers' information. The intent of the electronic interface was to provide resellers with online, direct access to service order entry and processing systems, number administration and resource systems, and trouble-reporting and line-testing systems. This would allow resellers to provide information to their customers on a real-time basis, just as RTC does with its customers. Thus far, Rochester has failed to deliver the system.

As a consequence, carriers—in particular AT&T—have complained that the initial service order process and general information flow have been totally inadequate. The "system" has been a progression of hand carrying, to fax, to e-mail. A direct consequence of this antiquated process is that the resellers are forced to provide inferior service to their customers, as compared with the service Rochester provides to its customers. At this juncture, AT&T is negotiating with Rochester on the next phase—the long-promised electronic interface. If this does not alleviate the situation, commission intervention will be required. This would be unfortunate because this issue is best resolved by a collaborative business solution.

LINK PRICING

In November 1992, the commission concluded that exchange access services of New York Telephone should be unbundled into two separate components: the link and the port, each with its own price. In the OMP, Rochester's residence exchange service has been unbundled into the requisite link and port. However, the monthly prices were set at fully distributed cost levels: $14.45 for the link and $2.84 for a usage-sensitive port (there is no unbundled flat rate residential port). The $14.45 monthly link price contrasts to Rochester's bundled retail rate for flat rate residential service of $12.96 per month. As might be expected, some parties objected to $14.45 monthly link price, arguing that it impedes the development of facilities-based, residential exchange service competition.

The commission observed that the disposition of this issue in the plan did not fully resolve the problem of how to develop a competitive framework that accommodates the existing residential service pricing in a reasonable manner, but looked to our ongoing Competition II proceeding for the answer. Thus, the commission put Rochester on notice that modifications to the OMP might result from commission action in Competition II proceeding. We have recently taken a step in that direction. A new phase has been initiated to look at the factual issues concerning the cost of links, and to address related policy questions concerning the pricing of links.

RECIPROCAL COMPENSATION FOR LOCAL TRAFFIC

The OMP provides for reciprocal compensation between Rochester and other facilities-based local exchange carriers for the termination of local traffic. Under this arrangement, a facilities-based local provider would pay Rochester for local switching and transport for local traffic it delivers for termination to a customer served by Rochester. Conversely, Rochester would pay the facilities-based competitor the same prices for the same termination service. If the traffic is in balance (i.e., equal in both directions within a 10% tolerance band), no payments would be made by either entity for local transport. The charges specified in the plan are 1.1 cents for local switching and 1.1 cents for transport. All traffic originating and terminating within Rochester's operating territory is considered local. This compensation arrangement was the product of negotiations primarily between Rochester and Time Warner.

Allegations were raised during the proceeding that the level of the charges—2.2 cents per minute—could result in a price squeeze vis-á-vis serving business customers who have volume discounts available to them. More recently, some carriers who have yet to begin providing service in the Rochester market have complained that the 2.2 cents for terminating traffic is too high relative to the actual incremental cost of about 1 cent to provide this service. These carriers also observe that the commission has set a lower compensation level at cost in the recently executed regulatory plan for New York Telephone.

The commission directed us to closely monitor the development of competition in Rochester in general, and this aspect of the plan in particular, and to report whether further action is warranted. Currently, Time Warner is the only other facilities-based carrier operating in the market; it is not complaining, presumably, because it is operating under a scheme it engineered. Nevertheless, as others enter the market, we will assess the impact of the compensation arrangement on the development of local competition; if changes are warranted, they will be undertaken.

COMMITTEE ON STANDARDS AND COOPERATIVE PRACTICES

The plan provides for a Committee on Standards and Cooperative Practices, which is composed of carriers that are certified by the commission to provide telephone service in the Rochester market. The stated objective of the committee is to develop technical standards and cooperative practices to facilitate the seamless interconnection of all facilities-based providers serving the Rochester market, and to further the development of a competitive market for both resellers and facilities-based carriers. We have a significant interest in the success of this committee to cooperatively devise business solutions to what we view as business problems, rather than turning to us for regulatory solutions. What we see, so far, is that parties agree with the intent of the committee, and are generally satisfied with the process for dealing with issues that have long lead times and/or are relatively noncontroversial. We also hear that it is not a satisfactory process for issues in contest—in need of a quick resolution. There is also a perception that Rochester Telephone Corporation considers itself more equal than others on the committee (i.e., if it does not want to deal with a given issue, the company will table it). This is not good news.

But this is a learning process for all. In our view, the committee was not intended to be the only forum for resolution of all competitive issues, but rather a good place to start. If the committee participants do not perceive that Rochester is adequately dealing with their important issues in a timely manner, they need to bring them to us. The commission has an Open Network Architecture (ONA) process that is designed to report to the commission on competitive issues under contention among parties in 30 days. We have also developed a rapid response team to deal with critical issues that arise between competitors, such as out-of-service conditions that affect service to consumers. Each of these forums is evolving; it is expected that, with more experience, it will become clear which forum is appropriate for a given issue.

AN UNANTICIPATED OUTCOME

An outcome that was totally unanticipated involves certain customers recently attracted to AT&T. Not surprisingly, when AT&T entered the market, it targeted its existing customers (i.e., those who were already AT&T toll customers). After AT&T suspended its marketing activities, it continued to attract a significant number of customers, but this included many about whom the company knew nothing, i.e., they were not AT&T toll customers. A growing number of these accounts are turning delinquent. It is not entirely clear at this juncture what the situation is: It appears that some customers have been recently terminated by Rochester, while others may see an opportunity to avoid paying for telephone service, and still others may need to avail themselves of LifeLine services, but may not know about their availability. In any event, it is a situation that bears further scrutiny and a solution that takes into account the interests of all of the market participants, as well as the consumers.

PACE OF COMPETITION

Although we did not have a precise timetable in mind with respect to the growth of competition in the Rochester market, the actual results to date are disappointing. We anticipated a lot more activity—more in the way of facilities-based competition and less in the way of resale activity—and the experience has been not much activity and the opposite mix. During the first half of 1995, Time Warner was not active in the market because it was conducting a limited technical and marketing trial in a residential apartment complex. Based on the trial results, it is now actively marketing, but only to multiunit residential dwellings. However, its entry into the residential market in general is pending another technical trial for single-family homes. We had expected Time Warner to have a more significant presence in the market by now. However, we did not anticipate the interest in resale that has been developed by AT&T and Citizens, and that might even be further stimulated with changes in the wholesale rate structure.

Notwithstanding this relatively pessimistic status report, let me leave you with some final thoughts on the pace of competitive growth. Recall that when the barriers to competitive provision of terminal equipment were finally removed, it took considerably longer than 9 months to achieve significant competitive inroads. The evolution of the competitive interexchange toll markets was slow and contentious. Today, the thought of monopoly terminal equipment or long-distance markets is preposterous. But monopolists do not give up their monopolies easily, despite what they say. In fairness to them, there are also difficult issues to resolve, such as interconnection architectures, compensation, and number portability. These issues will require refinement to be fully supportive of the development of a competitive local exchange market. It will not be an easy process, but one day the thought of a single provider of dial-tone telephone service and a single provider of cable-TV service will be equally preposterous. It will be several years in the making and will require a lot of intercompany operational cooperation and close regulatory oversight. During the transition, I expect a substantial degree of "refereeing" will fall to state regulators, who are much more the agents of change than the stumbling blocks that we are usually accused of being.

Access and Interconnection Pricing: How Efficient is the "Efficient Component Pricing Rule"?

Nicholas Economides
Lawrence J. White
New York University

The question of how a monopolist owner of a bottleneck facility should set the price for access to the facility by an entrant or rival supplier of a complementary component continues to be an interesting question for theory and policy.[1] This question is often framed in terms of a regulated monopolist vis-à-vis an entrant or rival in an unregulated complementary activity; but the issue can also arise in the antitrust context of an unregulated "essential facility" monopolist that is vertically

---

1. *See* William J. Baumol, *Some Subtle Issues in Railroad Deregulation*, 10 INT. J. OF TRANS. ECON. 341 (1983); *Telecom Corporation of New Zealand and Others v. Clear Communications Ltd,* Privy Council, House of Lords, U.K. (1994), Curtis M. Grimm & Robert G. Harris, *Vertical Foreclosure in the Rail Freight Industry: Economic Analysis and Policy Implications*, 5 ICC PRACT. J. 508 (1983); Henry McFarland, Railroad Competitive Access: An Economic Analysis, mimeo (1985); Nicholas Economides & Glenn Woroch, Benefits and Pitfalls of Network Interconnection, mimeo (1992); WILLIAM J. BAUMOL & GREGORY SIDAK, TOWARD COMPETITION IN LOCAL TELEPHONY (1994); William J. Baumol & Gregory Sidak, *The Pricing of Inputs Sold to Competitors*, YALE J. REG. 171 (1994); Alfred E. Kahn & William E. Taylor, *The Pricing of Inputs Sold to Competitors: Comment*, 11 YALE J. REG. 225 (1994); Henry Ergas & Eric Ralph, Pricing Interconnection: Is the Baumol-Willig Rule the Answer? mimeo (1994); Mark Armstrong & Chris Doyle, Interconnection and the Effects of Entry, mimeo (1994); Jean-Jacques Laffont & Jean Tirole, *Access Pricing and Competition*, 38 EURO. ECON. REV. 1673 (1994); and Mark Armstrong & John Vickers, The Predatory Access Pricing Problem, mimeo (1994).

integrated into a complementary upstream or downstream activity in which one or more other producers are present (or may enter).[2]

As technological changes and legal-regulatory changes have created more opportunities for competition in activities that are complementary to a still-regulated bottleneck facility, the policy relevance of the access pricing question has been heightened. Familiar examples include:

> Local telephone service entrants who must route calls to and from the customers of the incumbent (bottleneck monopoly) provider through the incumbent's switches.

> Long distance telephone service providers who must access customers via the local (monopoly) switched network; this example extends immediately to other providers of complementary telephone services. In these instances, the local monopolist is usually also an actual or potential provider of the long distance and other complementary services.

> Generators of electricity who wish to sell to ultimate customer-users but who can reach those customers only through a local (monopoly) distribution network (and possibly also a monopoly transmission system); again the local monopoly distributor typically also owns generating facilities.

> Sellers of natural gas who similarly wish to sell to ultimate customer-users but who can reach those customers only through monopoly gas transmission pipelines and/or a monopoly local distribution network.

In addition, eased merger standards in some sectors (e.g., railroads) have created local monopoly bottlenecks that generate an access problem for competing firms that provide complementary components or services.

A widely discussed "rule" for the pricing of access to these bottleneck facilities was originally proposed by Baumol[3] and has recently been popularized by Baumol and Sidak;[4] it is frequently described as the "efficient component pricing rule" (ECPR), which is the terminology that we will use in our subsequent discussion. The ECPR states that the appropriate access charge by the bottleneck monopolist to the providers (actual or potential) of a complementary component or service,

---

2. For recent discussions of the essential facilities doctrine, see Gregory Werden, *The Law and Economics of the Essential Facilities Doctrine,* 32 ST. LOUIS UNIV. L. REV. 432 (1987); James Ratner, *Should There Be an Essential Facilities Doctrine?* 21 UNIV. OF CALIF., DAVIS, L. REV. 327 (1988); and David Reiffen & Andrew N. Kleit, *Terminal Railroad Revisited: Foreclosure of an Essential Facility or Simply Horizontal Monopoly?* 33 J. of L & Econ. 419 (1990).

3. *See* Baumol, *supra* note 1.

4. *See* Baumol & Sidak, *supra* note 1, and BAUMOL & SIDAK, *supra* note 1.

which the monopolist also produces (and thus the other providers are rivals to the monopolist), is a fee equal to monopolist's opportunity costs of providing the access, *including* any forgone revenues from a concomitant reduction in the monopolist's sales of the complementary component.

The ECPR has a seductive logic: It ensures that a rival producer of the complementary component can provide service only if that producer is at least as efficient as the monopolist in the production of the complementary component; i.e., the ECPR ensures that production will not be diverted to an inefficient producer.

It is now well established that the ECPR holds as a first-best pricing principle only if a stringent set of assumptions holds:[5] the monopolist's price for the complementary service has been based on a marginal-cost pricing rule; the monopolist's and rival producer's components are perfect substitutes; the production technology of the component experiences constant returns to scale; the rival producer has no market power; and the monopolist's marginal cost of production of the component can be accurately observed.

In this paper, we will examine the consequences of relaxing some of these assumptions. We focus special attention on the case where the monopolist has been charging a price for the complementary component that is above all relevant marginal costs. As we show, in this case the ECPR's exclusion of inefficient rivals may be socially *harmful*; the market presence of even an inefficient rival could bring net social benefits, by causing the price to fall sufficiently so that the net gain to consumers (the reduction in the deadweight loss "triangle") would exceed the inefficiency costs of the rival's production.

Figure 4.1
Interconnection to a Bottleneck
The links owned by the monopolist are drawn in bold.

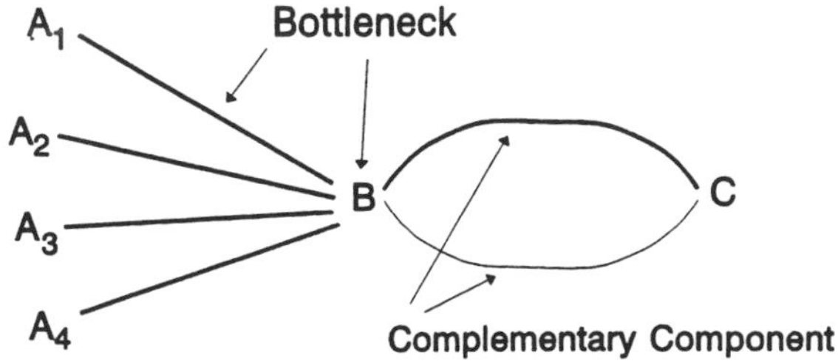

---

5. *See* Lafont & Tirole, *supra* note 1.

To help readers with the analysis that follows, we offer Figure 4.1 as a schematic of the framework that we are presenting. We describe our framework in terms of telephone services (but the other examples mentioned above are easily applied): The monopolist owns the switch at location B and provides local telephone service between and among customers at points A_1, A_2, A_3, etc. All of the local customers must use the monopolist's switch to complete (connect) their local calls and to gain access to other (complementary) services, such as long distance.

The same firm that provides the monopoly local service also provides service from points A_1, A_2, A_3, etc., through switch B to point C. This service could be "long distance"; or it could be additional "local" service to additional customers; or it could be some other complementary service (e.g., access to an information database) that requires the use of switch B. We will describe this ABC (or CBA[6]) service simply as "through service." There is at least one other potential or actual provider of the service to point C. This rival requires access to (through) the switch in order to provide the A_1, A_2, A_3, etc., customers with the "through" service ABC (and CBA). We assume, however, that the rival owns only facilities between BC, while the monopolist owns the switch B as well as its own facilities between BC (drawn in bold) and the links A1B, A2B, etc. In the language of the ECPR, the switch B is the monopoly bottleneck[7] and segment BC is the complementary component.

The early sections of the paper will assume the following:

The monopolist and the rival offer identical service over segment BC.

Service between B and C has value only as part of the through service ABC or CBA.

Constant returns to scale production technology applies to the production of service between B and C.[8]

The monopolist is able to charge prices for local service to its local customers that are sufficient to cover all of its costs of providing that local service (i.e., sufficient to cover the costs of providing links AB and the costs of the switch).

---

6. For a discussion of one-way networks and two-way networks, see Nicholas Economides & Lawrence J. White, *Networks and Compatibility: Implications for Antitrust*, 38 EURO. ECON. REV. 651 (1994). In two-way networks, such as telephone or rail systems, ABC and CBA are distinct goods or services. In a one-way network (such as an electricity grid), only one of these combinations is meaningful.

7. The segments *between* B and A1, A2, A3. etc., may or may not also be part of the monopoly bottleneck.

8. We will make the standard economics assumption that normal, competitive profit levels will be a component of the cost concepts discussed later in the article.

The price of the through service ABC (and CBA) is not subject to direct price regulation.

The consumer demands for local service and through service exhibit normal properties; e.g., the prices that consumers are willing to pay are indicative of the welfare or satisfaction that they receive from the services; at lower prices consumers want to buy more of the services, etc.

In later sections of the paper we will explore the consequences of modifying some of these assumptions.

The remainder of this paper will proceed as follows: Section II will lay out the structure of the basic ECPR and explore its logic. Section III will examine the consequences of the monopolist's price embodying a monopoly overcharge. Section IV will analyze the consequences of the components' not being perfect substitutes. Section V discusses the case when economies of scale are present. And Section VI will offer a brief conclusion.

THE LOGIC OF THE EFFICIENT COMPONENT PRICING RULE (ECPR)

The logic of the ECPR is readily demonstrated through a simple numerical example:

Suppose that the monopolist charges a price of $0.10 for through service ABC (or CBA). Suppose further that the monopolist's marginal costs of providing this service are $0.02 for segment BC and $0.05 for segment AB (including the relevant marginal costs of the switch B). *The EPCR simply states that the appropriate price or fee for the monopolist to charge to the rival for access to switch B (and for providing the connecting service AB) is $0.08:* The $0.05 of marginal costs relevant to segment AB plus the forgone net revenue of $0.03 that the monopolist loses when the rival provides the through service in lieu of the monopolist.

If the rival is being charged a fee of $0.08 for access, and the monopolist is charging $0.10 as its price to customers for through service, then the rival will be able to offer through service without incurring losses only if its marginal costs for segment BC are at or below $0.02—i.e., at or below the marginal costs (over BC) of the monopolist.

Thus, the ECPR ensures that the rival enters and produces in the market only if its costs are no greater than those of the monopolist; inefficient diversion of production away from the monopolist will not occur as a consequence of the presence of the rival in the market. Further, if all of the conditions mentioned in the Introduction (including pricing by the monopolist at marginal cost) are satisfied, the ECPR will provide global efficiency.

THE MONOPOLIST INITIALLY HAS MARKET POWER IN THE
COMPLEMENTARY COMPONENT

The previous section made no explicit assumption as to the basis for the monopolist's price for the through service. We now explicitly assume that in the absence
of the rival the monopolist is able to charge the full profit-maximizing monopoly
price for the through service.[9] In turn, this maximizing behavior implies that the
monopolist's markup over marginal costs is directly related to those marginal
costs and to the elasticity of demand for the service.

Figure 4.2
Monopoly Pricing and Deadweight Loss

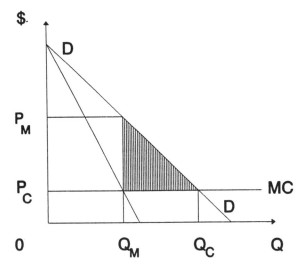

This monopoly outcome can be portrayed in a familiar geometric diagram, provided in Figure 4.2. With demand curve (DD) for through service and constant marginal
costs (MC) for the monopolist, the profit-maximizing monopolist would charge a
price of P_M and produce a quantity Q_M. (This can be compared to the price PC and
quantity QC that would be yielded by a marginal-cost pricing regime.) The monopolist's markup over marginal costs is the vertical distance between P_M and MC.

9. Baumol & Sidak, *supra* note 1, and BAUMOL & SIDAK, *supra* note 1, briefly acknowledge the
possibility that the monopolist's price for the complementary component (or, in our case, through service) might reflect market power and not be based on a marginal-cost pricing rule, and they acknowledge that this possibility would mean that the optimal properties of the ECPR would not hold; but they
nevertheless devote virtually all of their analysis and discussion to the case where market power is absent. Kahn & Taylor, *supra* note 1, note this brief acknowledgement and the possibilities that might
follow from market power, but they do not fully pursue the point.

The ECPR described in the previous section would prescribe that the monopolist's markup or overcharge be a component of the access fee. Thus, the EPCR would deter inefficient rivals (those with marginal costs that are higher than the monopolist's MC) and prevent inefficient production. *But the ECPR also protects the monopolist from any competitive challenge by these rivals and thus protects the monopolist's profits; and the ECPR preserves the allocative or consumption inefficiency that results from the monopolist's excessively high price for through service.* The social loss of the protected monopoly in usually calculated as the "deadweight loss triangle" (DWL) shown as the shaded area in Figure 4.2: The loss of consumers' surplus by demanders who are shut out of the market by the monopolist's high price.

It is easy to see that, if the monopolist has market power in the market for the complementary component, the ECPR is the monopolist's profit-maximizing access fee when the rival is less efficient that the monopolist, since the ECPR precludes entry and allows the monopolist to continue to reap its full monopoly rents.[10]

Production by a Less Efficient Rival Could Yield Net Social Gains

If the monopolist were required (e.g., by regulation) to levy a lower access fee, a less efficient rival could begin production. But it is nevertheless possible that social welfare would increase, because the diminished DWL from the lower price that could accompany entry could more than compensate for the social cost of the rival's inefficient production.[11] The magnitudes of the price decrease and the rival's inefficiency, and also the fractions of post-entry production that the rival captures, will be crucial to this determination.

BERTRAND COMPETITION. To show that entry by even an inefficient rival could yield socially beneficial results, we assume that the monopolist is restricted (e.g., by regulation) to levying an access fee that is equal only to the actual marginal costs of access (i.e., the marginal costs of segment AB, including the switch). We further assume that an entrant to segment BC has higher costs than the monopolist by an amount t, where $0 \leq t \leq (P_M - MC)$.

Suppose that in response to the prospect or actuality of entry, the monopolist practices limit pricing: It sets the price (P_J) at which both producers sell at a level that is just equal to the entrant's costs. Under this pricing regime the entrant will capture some share θ ($0 \leq \theta \leq 1.0$) of the joint market sales (Q_J) of through service.

---

10. The profit-maximizing access fee when the rival is more efficient than the monopolist is discussed below in the text.

11. The analysis that follows in the text is an adaptation, in reverse form, of the approach of Oliver E. Williamson, *Economics as an Antitrust Defense: The Welfare Tradeoffs*, 58 AM. ECON. REV. 18 (1968).

Figure 4.3
Dead Weight Gain and Production Inefficiency.

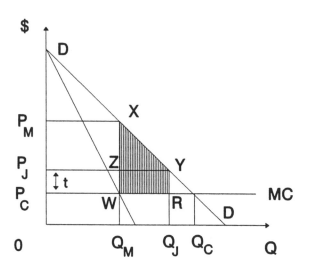

In the terms of Figure 4.3, consider the change from pure monopoly (point X) to limit pricing (at point Y). The approximate social gain from this change is a combination of a "triangle" gain in consumers' surplus ($\frac{1}{2}\Delta P \cdot \Delta Q = \frac{1}{2}[P_M - P_J]\cdot[Q_J - Q_M]$), represented by area XYZ, and a "rectangle" gain in producers' surplus ($t\cdot\Delta Q = t\cdot[Q_J - Q_M]$), represented by area ZYRW; the combined areas are the trapezoid XYRW. This gain is offset by the production inefficiency: the entrant's cost disadvantage multiplied by the entrant's production volume, or $t\theta Q_J$, which represents some fraction (θ) of the rectangle P_JYRP$_C$. The net gain can be either positive or negative, depending on the values of t, θ, and the elasticity of demand (e).

For specified values of θ and e, it is possible to find (solve for) a maximum level of cost disadvantage (t*) that just eliminates the net social gain.[12] Then, for the specified values of θ and e, entry by a firm with cost disadvantage t, t < t*, will yield a social gain. In Table 1 we provide the t* values associated with a range of plausible values for e and for some benchmark values of θ: $\theta = 0$ (the mere threat of entry causes the monopolist to practice limit pricing); $\theta = 0.5$ (a Bertrand equal-sharing of the market between the monopolist and the rival); and $\theta = 1.0$ (the "worst" case, in which the less efficient rival somehow captures the entire market). As can be seen, for these "realistic" elasticity values, a rival may experience a substantial cost disadvantage and still be the vehicle for a net social welfare improvement. These results are even more striking if the rival's cost disadvantage is

12. Appendix A provides the calculations that yield t*.

expressed as a percentage of the monopoly's profit maximizing price differential (or profit margin).

TABLE 4.1
Matrix of t* Values for Bertrand Competition

| | Elasticity e | | |
θ	2.0	3.0	4.0
0	1.00ᵃ (100%)	0.50* (100%)	0.33ᵃ (100%)
0.50	0.67 (67%)	0.33 (65%)	0.21 (65%)
1.0	0.50 (50%)	0.24 (47%)	0.15 (46%)

* Profit maximizing monopoly markup over marginal cost (profit margin).

Note: The numbers in parenthesis represents t* expressed as a percentage of the maximum monopoly markup.

COURNOT DUOPOLY.   We have thus far performed our analysis on the basis of an assumption of a "tough" pricing environment between the monopolist and the rival: Bertrand-like limit pricing by the monopolist. More accommodating behavior between the monopolist and the rival would yield higher prices for any t by the rival and hence a lesser likelihood of a social welfare gain; at the limit, if both the monopolist and rival sell at the monopoly price (P_M), with the monopolist ceding to the entrant some share q of the monopoly volume (Q_M), then the outcome yields only the social loss ($t\theta Q_M$) due to the inefficient production of the entrant.

As an example of an intermediate level of pricing toughness, we assume that the competition between the incumbent and the entrant is à-la-Cournot; i.e., each producer adjusts its own production quantity on the assumption that the other producer's quantity will remain unchanged. If a linear demand curve is used for illustration, then, similar to our Bertrand demonstration, we can solve for a t*, such that all t < t* will yield a net social gain.[13] For this Cournot conjecture by the producers, the market share of the entrant (θ) is endogenous, with $\theta \simeq 0.31$ (i.e., a market share of 31%) at t = t* and $\theta = 0.50$ at t = 0. Table 2 shows the critical values of t* for the same "realistic" values of e that were provided in our Bertrand case. As can be seen in a comparison of Table 2 with Table 1, the t* values for this Cournot case are smaller than for the Bertrand case, because Cournot competition implies higher prices and a substantial market share for the less efficient rival. Still, again, *entry by a firm with a non-trivial production cost disadvantage relative to the monopolist can result in a net social gain.*

---

13. Appendix B shows the steps that are required to solve the Cournot competition problem and generate a solution for t*.

TABLE 4.2
Values of t* for Cournot Competition

	Elasticity e	
2.0	3.0	4.0
0.29 (29%)	0.17 (33%)	0.12 (35%)

Note: The numbers in parentheses represents t* expressed as a percentage of the maximum monopoly markup.

We also note that if more than one rival is present in the market, so that there is competition among a number of firms, the equilibrium price is likely to be closer to the Bertrand limit pricing case, and definitely lower than for the Cournot duopoly. Thus, in the presence of more than one rival, the welfare gain from the competition that they bring is more likely to be positive.

A SUMMING UP. The ECPR's apparent strength—its exclusion of an inefficient rival—may also be its drawback when the monopolist is charging high prices (in excess of all relevant marginal costs) for the complementary component. In that case, the exclusion of the rival also prevents the possibility of a lower price for the complementary component, with its attendant reduction in consumers' deadweight loss; the net social gain from the rival's presence could be positive. The extent of the price decrease, the size of the rival's cost advantage, and the rival's equilibrium market share are the key determinants of whether the rival's presence in the market would be beneficial. In turn, the oligopolistic conjecture held by the two producers and the market elasticity of demand will influence these magnitudes. At one extreme, if the monopolist practices Bertrand-like limit pricing, the presence of a rival with even a substantial cost disadvantage can be socially beneficial. At the other extreme, if the monopolist and the rival jointly maintain the monopolist's previous high price (with the monopolist simply ceding some market share to the rival), then any market presence by a less efficient rival will be socially deleterious. Cournot duopoly yields an equilibrium price that is between these extremes, with an endogenously determined market share for the rival. Even with Cournot duopoly, nontrivial cost inefficiencies by the rival are consistent with a net social gain.

Further reflection on the ECPR reveals both a greater universality to its logic but also a greater universality to the critique that we have just offered.[14] Though the ECPR is usually presented in terms of access to a bottleneck facility, its logic extends to any entry by any rival firm into any market where there is an incumbent. If the sole goal of social policy is to prevent inefficient production by an entrant from displacing more efficient production by the incumbent, then the ECPR principle—the entrant must reimburse incumbent for the latter's opportunity costs, in-

---

14. We are grateful to Timothy Brennan for pointing this out to us.

cluding forgone net revenues—will achieve that goal. But, *if the incumbent is exercising market power, then the use of the ECPR will also protect the incumbent and preserve its market power against the competitive erosion of prices and margins that even less efficient rivals could bring.* If the ECPR is placed in this context, the lustre of its rationale tarnishes rapidly.

An "entry tax" that required market entrants generally to reimburse incumbents for their forgone net revenues would quickly be seen as a protective and anti-competitive device and would likely receive little support from policy-oriented economists, despite any claims that the tax would preclude inefficient production. The ECPR is just this type of entry tax.

The ECPR's Harm When the Rival is More Efficient Than the Monopolist

The previous section only addressed instances where the rival is less efficient than the monopolist in the production of the complementary component. Even in those circumstances, as we demonstrated, the net effect of the ECPR could be harmful.

If the rival is more efficient than the monopolist, then the monopolist's profit maximizing strategy would generally[15] be to cede production of the complementary component to the more efficient rival and to reap its monopoly profits through an appropriate access fee.[16] This access fee could be either greater or less than the ECPR, depending on the shape of the demand curve for through service.

Thus, for this case the presence of the ECPR would allow production to shift to the more efficient producer. But the ECPR access fee would mean that consumers would continue to endure the inefficiencies of the artificially high price for through service. And the divergence of the ECPR from the monopolist's profit-maximizing access fee could imply either further distortion (if the ECPR is higher than the profit-maximizing fee) or a lessening of distortion (if the ECPR is lower).[17]

THE COMPLEMENTARY COMPONENTS OF THE MONOPOLIST AND THE RIVAL ARE IMPERFECT SUBSTITUTES

We now assume that the complementary components of the monopolist and the rival are imperfect substitutes; i.e., the monopolist and the rival compete in offering through service, but their service offerings are not identical. To continue with our tele-

---

15. We discuss exceptions below in the text.

16. By ceding all production of the complementary component to the rival, the monopolist might be creating problems of vertical supplier-customer relationships, with the consequent problems of double marginalization, etc. But the monopolist's ability to self-supply the complementary component at its own marginal cost would put a limit on the extent to which the rival could attempt to exploit that position.

17. The discussion in this section thus indicates that a more complete "Williamson" type analysis should also include the possibility that more efficient rival might enter, and the absence of the ECPR would mean that consumers would enjoy the full benefits of the lower price brought by the entrant.

phone example, the rival's through service (e.g., long distance service) might involve faster connections or a higher percentage of completed calls, but also an increased level of static on the line, as compared with the monopolist's long-distance service.

The introduction of imperfect substitutes immediately calls into question the meaning of any comparisons of production cost efficiency between the two components. Since the two components now have different attributes and satisfy (somewhat) different demands, comparisons of their unit costs have little or no meaning. It is rarely interesting or analytically worthwhile to compare the "unit" costs of an apple producer and an orange producer.

Even if the units of the two complementary components are somehow comparable, the analog of the ECPR's consequences for the perfect substitutes case is that the imposition of the ECPR to the rival's imperfect substitute would exclude a rival's production entirely when there was no customer with a willingness to pay for the rival's service such that the rival's price-less-costs margin could exceed the ECPR access fee. Unless the ECPR were based on a marginal-cost pricing rule (and thus in this case there were no customers of the rival's service whose willingness to pay would cover all relevant marginal costs), the exclusion of these "modest" willingness-to-pay customers by the ECPR access fee would not serve the goal of promoting production efficiency.[18]

If we move away from a production efficiency criterion, then an access fee might serve some other purpose—say, maximize the monopolist's profits, or help solve a Ramsey pricing problem if inadequate revenues can be earned from the monopolist's bottleneck service customers. Only by chance would the ECPR access fee be the solution to either of these problems.

ECONOMIES OF SCALE

In the presence of economies of scale in the production of the bottleneck service and/or the complementary component, the ECPR is again unlikely to provide a first-best pricing outcome.

Ramsey Pricing

Suppose, contrary to our earlier assumption, that the monopolist is unable to earn sufficient revenue in the bottleneck market to cover its costs (e.g., because of economies of scale). In the absence of any other source of funds, the regulator must extract a contribution from the customers of through service. To maximize social welfare, the regulator must solve a Ramsey problem: select the set of prices for local service and through service that maximizes consumers' surplus while also

---

18. *See also* Laffont and Tirole, *supra* note 1; Armstrong and Doyle, *supra* note 1; and Armstrong and Vickers, *supra* note 1.

covering the costs of those services. The resulting Ramsey price for through service would involve, in essence, an excise tax that is levied on through service. If the monopolist is the low cost producer of the complementary component, the monopolist "pays the excise tax to itself;" if the rival is the low cost producer, the rival pays the tax to the monopolist as an access fee. To ensure optimality, the regulator would have to regulate directly the excise tax and the resulting price of through service. Only by chance would the ECPR access fee be identical to the Ramsey excise tax.[19]

The Monopolist May Use the ECPR to Exclude a More Efficient Rival

There are at least two circumstances in which the monopolist's profit-maximizing strategy is to exclude the rival, even when the rival is more efficient at producing the complementary component, rather than cede production to the rival. In both instances, the monopolist would find worthwhile to understate its own marginal costs of production of the complementary component and then impose a heightened ECPR on the rival.

In the first case, we now assume either that segment BC is a separate stand-alone market (as would be true for local railroad freight hauling) or that it is a complementary component for through service to (or from) points D, E, F, etc., that are served by other firms (but not by the original monopolist). We also depart from our assumption of constant returns to scale in the production of the complementary component and instead assume that there are economies of scale. If the monopolist can exclude the (more efficient) rival from offering through service in the ABC market, this could sufficiently deprive the rival of the benefits of scale so that the (less efficient) monopolist would also be the monopolist of the BC segment and reap monopoly profits from the stand-alone BC service or from the through service to the other points. To achieve this outcome, the monopolist could understate its marginal costs of production of the complementary component (BC service) and then employ the ECPR criterion to levy an exclusionary access fee vis-à-vis the rival.

As a second case, we alter our earlier assumption about the monopolist's behavior in the bottleneck market. We assume that the monopolist is constrained by regulation to earn zero excess profits in the bottleneck market. If, however, the regulator cannot observe the monopolist's costs perfectly, then the monopolist can increase its aggregate profits by claiming that some of its costs of production of the complementary component should be treated (for regulatory purposes) as

---

19. *See* Lafont and Tirole, *supra* note 1; Armstrong and Doyle, *supra* note 1; and Armstrong and Vickers, *supra* note 1.

costs of production of the bottleneck services.[20] To the extent that the monopolist succeeds in this "creative accounting" for some of the marginal costs of production of the complementary component, these apparently lower marginal costs would justify a higher ECPR access fee. This higher access fee will again deter some efficient rivals—i.e., those with marginal production cost levels that are less than the monopolist's true marginal costs of production of the complementary component but greater than the remaining marginal costs that the monopolist still attributes to the production of the complementary component.[21]

CONCLUSION

The "efficient component pricing rule" (ECPR) for access pricing has a seductive logic: It appears to ensure that only efficient production of a complementary component (to a monopoly bottleneck service) will occur. The ECPR holds as a first-best principle, however, only under a stringent set of assumptions.

Only empirical observation can ascertain whether these assumptions are closely enough approximated in reality that the ECPR is a reasonable basis for policy. Our professional judgment is that real-world conditions are often likely to diverge importantly from the necessary assumptions. We are most concerned about the assumption that the monopolist's pricing of the complementary component is driven by marginal cost (or Ramsey) principles. If, instead, the monopolist's price reflects the exercise of market power, then the ECPR will protect that market power and prevent consumers from benefitting from the price competition that a rival (entrant) could bring. We show that there are quite reasonable circumstances under which the presence of even a less efficient rival would bring about a positive net social welfare change. We also explore the consequences of the loosening of some of the other assumptions.

In sum, in real-world settings policy makers should be wary of blind devotion to the ECPR. It has dangers as well as benefits, and the real-world settings may well be ones in which the dangers outweigh the benefits.

---

20. This is a well known possibility and was one of the major arguments for separating regulated monopoly local telephone service from other complementary services. *See, for example,* Roger G. Noll & Bruce M. Owen, *The Anticompetitive Uses of Regulation: United States v. AT&T,* in THE ANTITRUST REVOLUTION: THE ROLE OF ECONOMICS 328 (John E. Kwoka & Lawrence J. White eds., 1994); Timothy J. Brennan, *Why Regulated Firms Should Be Kept out of Unregulated Markets: Understanding the Divestiture in United States v. AT&T,* 32 ANTITRUST BULL. 741 (1987); and Timothy J. Brennan, *Cross-Subsidization and Cost Misallocation by Regulated Monopolists,* 2 J. REG. ECON. 37 (1990).

21. Though this incentive for cost shifting (which is, in essence, regulatory evasion) is not dependent on the presence of the ECPR, the ECPR may well add the distorting element of the exclusion of an efficient rival.

APPENDIX A

In this appendix, we show how the net social gain can be expressed in terms of t, θ, and e, and how t* can subsequently be determined. We assume that the incumbent has marginal costs of c and that the rival has marginal costs of c + t.

If the incumbent can exercise market power and act as a monopolist, its price (P_M) is

$$P_M = c \cdot e/(e - 1).$$

If, in the presence of the rival, the monopolist practices limit pricing, the price (P_J) at which both firms will sell $(P_J = P_M = P_R)$ is

$$P_J = c + t.$$

The consumers' surplus "triangle" gain from this switch from monopoly to limit pricing is

$$
\begin{aligned}
\tfrac{1}{2} (\Delta Q)(\Delta P) \quad &= \tfrac{1}{2} (Q_J - Q_M) \cdot (P_M - P_J) \\
&= eQ_J(\Delta P)^2/(2P_J) \\
&= eQ_J[c/(e - 1) - t]^2/[2(c + t)].
\end{aligned}
$$

The producers' surplus "rectangle" gain is

$$
\begin{aligned}
t \cdot (\Delta Q) \quad &= t \cdot (Q_J - Q_M) \\
&= te(\Delta P)Q_J/P_J \\
&= teQ_J[c/(e - 1) - t]/[2(c + t)].
\end{aligned}
$$

Finally, the production inefficiency loss "rectangle" is

$$t\theta Q_J.$$

If the last expression is subtracted from the sum of the preceding two expressions, the net social gain that results from entry and limit pricing is

$$Q_J[e\{[c/(e - 1) - t]^2 + t[c/(e - 1) - t]\}/[2(c + t)] - t\theta].$$

If we normalize by setting c = 1, we find that this expression is positive if and only if

$$t < t* = \frac{\theta - \sqrt{\theta^2 + \dfrac{e^2 + 2\theta e}{(e - 1)^2}}}{-(e + 2\theta)}.$$

It is this expression that is the basis for the values in Table 4.1.

APPENDIX B

In this appendix, we provide the solution to the simple Cournot game and the subsequent determination of t*, where the incumbent has marginal costs of c, the rival has marginal costs of c + t, and the inverse demand for the service is linear, $P = a - bQ$, expressing the relationship between price (P) and quantity (Q). Quantity, in turn, is the sum of the incumbent's production (Q_I) and the rival's production (Q_R); the price at which they sell jointly is P_J.

The incumbent's profits are

$$\Pi_I = [a - b(Q_I + Q_R)] \cdot Q_I - cQ_I.$$

The rival's profits are

$$\Pi_R = [a - b(Q_I + Q_R)] \cdot Q_R - (c + t) \cdot Q_R.$$

The non-cooperative equilibrium of the quantity setting (Cournot) game is

$$Q_I = (a - c + t)/(3b), \quad Q_R = (a - c - 2t)/(3b),$$
$$Q_J = Q_I + Q_R = [2(a - c) - t]/(3b),$$
$$P_J = (a + 2c + t)/3.$$

Given the quantities determined at the equilibrium, the market share of the rival is

$$\theta = Q_R/Q_J = [a - c - 2t]/[2(a - c) - t].$$

Since the pure monopoly equilibrium for the incumbent is

$$Q_M = (a - c)/(2b), \quad P_M = (a + c)/2,$$

the consumers' surplus "triangle" gain from the switch from the monopoly to Cournot duopoly is

$$^1/_2 (\Delta Q)(\Delta P) = {}^1/_2 (Q_J - Q_M)(P_M - P_J) = (a - c - 2t)^2/(72b).$$

The producers' surplus "rectangle" gain from the switch is

$$(Q_J - Q_M)(P_J - c) = [a - c - 2t][a - c + t]/(18b).$$

Finally, the production inefficiency loss "rectangle" attributable to the rival's output is

$$tQ_R = t(a - c - 2t)/(3b).$$

If the last expression is subtracted from the sum of the preceding two expressions, the net social gain that results from entry under Cournot behavior is

$$[a - c - 2t][5(a - c) - 22t]/(72b).$$

This expression is positive if and only if both bracketed terms are positive (or both are negative). The economically relevant case requires

$$5(a - c) - 22t > 0,$$

or, equivalently,

$$t < t^* = 5(a - c)/22.$$

The elasticity of demand at the Cournot equilibrium is

$$e = (\Delta Q/\Delta P)/(Q_J/P_J) = [a + 2c + t]/[2(a - c) - t].$$

If we normalize by setting $c = 1$, this last expression can be solved for a in terms of e and t, and this result can then be used to solve for t^* in terms of the elasticity e:

$$t^* = 5/(13e - 9).$$

It is this expression that provides the basis for the values in Table 4.2.

ACKNOWLEDGMENTS

This chapter was originally published in the fall 1995 issue of the Antitrust Bulletin; permission to reproduce it here is gratefully acknowledged. The authors thank Timothy Brennan for helpful comments on an earlier draft of this article. The authors also thank the participants at the CEPR/CREST-LEI conference on Mobile Telephony, the "Utilities Regulation Network" conference at the Catholic University of Milan, and of the 1995 Annual National Conference of Economic Research in France for helpful comments and suggestions.

II

PRICING PROBLEMS IN COMPLEX NETWORKS

5

Pricing of ATM Network Services

Qiong Wang
Carnegie Mellon University

Marvin A. Sirbu
Carnegie Mellon University

Jon M. Peha
Carnegie Mellon University

Different communication networks have been developed to meet the specific needs of different services, including telephone networks for voice, cable-TV networks for video, and the Internet for computer communications. With the emergence of technology to support cell switching based on asynchronous transfer mode (ATM), it is clear that ATM integrated-services networks are going to replace today's single-service networks, and become the networking standard for the telecommunications industry in the near future. In ATM networks, all types of information, such as video, voice, image, and data, are organized as streams of fixed-length packets (cells). By controlling the average and variance of transmission speed of each cell stream, different services can be provided to satisfy consumers different needs.

Before ATM network services can be efficiently provided, the pricing issue should be carefully addressed. Developing a general optimal pricing mechanism for any given ATM network is a broad and interesting problem. The result not only depends on economic parameters, such as consumer demand, but also is affected by technical specifics, such as network topology and routing algorithms. Although this chapter cannot consider every detail at the same time, it focuses on pricing ATM network services on point-to-point links. There are reasons to believe that the development of ATM networks will start with single links, so it is appropriate to solve the link pricing problem first. The solution sheds some light on the more general networking pricing problem.

The goal of the pricing decision is to maximize social welfare if the network is controlled by a regulator, or to maximize profit if the network is operated by an unregulated natural monopoly. In either case, one should decide the price of a service based on the cost of providing it. However, unlike that of providing a tangible product, the physical cost of providing a network service is usually negligible once the infrastructure is in place. The major cost that needs to be considered in the pricing decision is the opportunity cost (i.e., the cost of denying services to some customers to serve others).[1] Unless the network capacity exceeds potential consumer demand (which is unlikely to be the case for ATM networks), the opportunity cost can be very significant, and it should be quantified and incorporated into the pricing mechanism. In ATM networks, the network capacity is shared by different services. In quantifying the opportunity cost for each service, one needs to consider the following service characteristics.

Performance Guarantee

Different ATM network services offer different levels of performance to different traffic streams, where performance is characterized by the distribution of end-to-end delay experienced by transmitted cells.[2] For example, in offering services for voice communications, the network can tolerate a maximum delay of 50 ms when using a split-echo suppressor. Depending on encoding techniques, the probability of exceeding delay deadline can be as great as 1%–20% without significant degradation, whereas for high-definition television (HDTV), the maximum tolerable delay might be close to 10–20 ms, and the percentage of cells delayed beyond that threshold should be considerably lower [1].

Some services, such as video and voice, have no value to users unless some specific performance objectives are guaranteed to be met during a call. These services are usually provided as *guaranteed services*, which means the network makes a commitment to meet performance objectives. For other services, like file transfer and e-mail, although a lower delay is preferred, the service will always have some value to users even if the network does not meet expected performance objectives [2]. The network can provide these services as "best-effort" services, which means no performance is guaranteed.

To make performance guarantees, the network must ensure that a certain amount of capacity is available for guaranteed traffic when needed. The amount of such

---

1. If the network operator's goal is to maximize profit, the opportunity cost is the profit foregone by excluding some users from services. If the goal is to maximize social welfare, the opportunity cost is the welfare foregone.

2. Cell loss can be viewed as a special case of cell delay time; the lost cell is considered to experience an infinite delay.

needed capacity increases with the number of calls in progress using guaranteed services. This is referred to as *reserved capacity*, although any unused capacity (regardless of whether it is reserved) can be used to provide best-effort service [3]. Once the amount of reserved capacity reaches a given threshold, the network will reject new requests for guaranteed services. Best-effort service can be interrupted whenever there is not enough capacity, and it can be resumed later when capacity becomes available [4]. Hence, the service does not need any capacity reservation. Consequently, a guaranteed service can tie up network resources by both reserving capacity (which will affect the provision of other guaranteed services) and using capacity (which will affect the provision of both guaranteed services and best-effort service), and a best-effort service can tie up network resources only by using the capacity. This difference should be reflected in an efficient pricing mechanism.

Consumption of Network Capacity

In an ATM network, calls of different services may insert different amounts of data into the network. For example, a voice call transmits data at tens of kilobits per second, whereas a HDTV call usually transmits at megabits per second. The amount of reserved capacity depends not only on average data rate, but also on the traffic pattern. For example, when only a few calls are in progress, services that generate bursty traffic (as in the case of compressed video delivery) will need to reserve more bandwidth than the average data rate to accommodate the peak transmission rate. The amount of reserved capacity approaches the average data rate only when there are many calls in progress, so the total traffic can be smoothed by multiplexing many bursty streams together. Everything else being equal, an efficient pricing algorithm should charge a higher price for services that reserve and/ or use more network capacity.

Time of Service

Because demand changes with time of day, the network capacity is more valuable at some times and less valuable at other times. If a consumer uses a network service at a time when the network capacity is more valuable, he or she should be charged more.

Service Duration

Everything else being equal, calls of longer duration consume more network resources, therefore they should be charged more. In addition, if a call lasts longer, as is likely to be the case for video on demand, price must reflect both the current value of network capacity and the value of network capacity throughout the duration of the call.

So far, no pricing approach that considers all of these factors has been implemented or proposed. The two most commonly used pricing schemata in practice are: (a) connection time-based pricing, which is used in long-distance telephone services; and (b) capacity-based pricing, which is independent of usage and is used in the Internet and leased-line services. In the connection time-based pricing, the price is a function of service duration and time of service; the amount of network capacity reserved or used within each unit of time is ignored. In capacity-based pricing, consumers are billed according to the maximum amount of network capacity they can use, regardless of how much capacity they actually use, how long, and when the capacity is used. Neither of the two pricing schemata is efficient for services like e-mail, file transfer, or telnet, which generate variable bit-rate traffic and require a different level of performance guarantee. A spot-price approach was proposed in recent years as an alternative for pricing the Internet [5]. In a spot-price mechanism, consumers are required to mark each packet with their willingness to pay. At each moment, the network sends packets with the highest willingness to pay, and charges a spot price that equals the lowest willingness to pay among the packets sent at that moment. The spot price will automatically be higher during peak hours and lower during off-peak hours, so it reflects different values of network capacity at different times of day. Because a spot price is charged for each packet sent, the more packets a user sends, the higher the total price he or she will be charged. Consequently, the average data rate is also considered in spot-price scheme. However, under a spot-price scheme, it is not possible to differentiate bursty calls from constant bit-rate calls, nor is it possible to provide a performance guarantee for a service, because any packet can be dropped if a consumer has not marked it with a willingness to pay high enough relative to other packets. As a result, a spot price may work with networks that offer only best-effort service, but not for integrated-services networks, which also offer guaranteed services.

This chapter develops a mechanism to price ATM network services, considering all factors discussed earlier. The next section presents the ATM network service model. Based on that service model, the optimal pricing framework is formulated, the resulting pricing schedule is discussed, and the advantage of that schedule is demonstrated.

NETWORK SERVICE MODEL

This section presents a single-link ATM network service model. It measures the capacity of that link in terms of the maximum number of cells that can be transmitted per unit of time, which is assumed to be a constant C_T. The following discusses the assumptions about network services, and then introduces two constraints on network capacity consumption. These constraints are used later to quantify opportunity costs for ATM network services in the optimal pricing model.

ATM Network Services

As mentioned earlier in the chapter, ATM network services can be classified as either guaranteed or best-effort services. The quantity of guaranteed services is measured by the number of calls completed, and that of best-effort service by the number of cells sent. In this pricing scheme, guaranteed services are divided into N classes. Two calls belong to the same class if and only if they have: (a) the same performance objectives; (b) the same traffic pattern, which implies both average cell transmission rate and burstiness are the same;[3] and (c) the same distribution of call duration.

For any class i guaranteed service ($i = 1,N$), its average cell transmission rate is denoted as s_i. The call duration is exponentially distributed with mean $1/r_i$, where r_i is the call departure rate. For each guaranteed service i ($i = 1,N$), call arrivals are Poisson distributed with an expected rate of λ_i—that is, within a very small time interval ($t, t + \Delta t$), the probability that there is no call arrival is $1 - \lambda_i \Delta t$, and that of one call arrival is $\lambda_i \Delta t$, and the probability that there is more than one call arrival is negligible.

We assume that $[\phi_1(t), \phi_2(t),..., \phi_N(t)]$ is the pricing schedule given by network operators, where $\phi_i(t)$ is the price of making a service i call for one unit of time at time t. Assuming that the exponentially distributed call duration is not affected by $\phi_i(t)$, consumers starting a service i call at time t are expected to pay an average price of:

$$p_i(t) = \int_t^{+\infty} \phi_i(\tau)e^{-r_i(\tau-t)}d\tau \tag{1}$$

$p_i(t)$ can be uniquely determined by $\phi_i(t)$ as long as $\phi_i(+\infty)$ is finite. By differentiating both sides of Equation 1 with respect to t, $\phi_i(t)$ can also be uniquely determined by $p_b(t)$. Therefore, to simplify our analysis, we assume that network operators manipulate $[p_1(t), p_2(t),..., p_N(t)]$ to maximize their utility. We assume for any service i, call arrival rate at time t is a function of $p_i(t)$ and time of day (i.e., $\lambda_i = \lambda_i[p_i(t),t]$, $i = 1,N$).

We also assume that the network only provides a single class of best-effort service, and requests consumers to reveal their willingness to pay for sending each cell in a given stream during the connection establishment process. We assume network operators adopt a spot-pricing scheme, by which a spot price, $p_b(t)$, will be set at each time t. During each small time interval ($t, t + \Delta t$), only cells for which

---

3. We assume the traffic of a guaranteed service can be modeled as a stochastic process $x(t)$, where $x(t)$ is the number of cells that need to be transmitted. Two calls have the same traffic pattern if both of them can be described by $x(t)$.

consumer willingness to pay is higher than $p_b(t)$ will be admitted into the buffer; other cells will be dropped. For any given $p_b(t)$, let $\lambda_b[p_b(t),t]$, be the resulting average cell admission rate (i.e., the rate of admitting cells into the buffer), which is a function of both spot price and time of day.

Constraints on Network Capacity Consumption

To meet performance guarantees, in ATM networks, admission control is enforced for all guaranteed services. Networks block such calls when there is not enough capacity to guarantee adequate performance. The desired call blocking rate for each service is usually specified during the network capacity planning process. Because call arrival rate fluctuates with time of day, the desired blocking rate may also change over time. This chapter assumes that the desired call blocking rate, denoted by $\overline{\beta}_i(t)$ ($i=1,N$), is exogenously given. Interested readers are referred to [6] for a discussion on how to select $\overline{\beta}_i(t)$.

For each guaranteed service i, let $q_i(t)$ be the expected number of calls in progress at time t. Given $\lambda_i[p_i(t),t]$ and r_i, $q_i(t)$ can be determined by:

$$\frac{dq_i(t)}{dt} = [1-\overline{\beta}_i(t)]\lambda_i[p_i(t),t] - r_i q_i(t) \tag{2}$$

Define $A[q_1(t), q_2(t),...,q_N(t), \overline{\beta}_1(t), \overline{\beta}_2(t),...,\overline{\beta}_N(t)]$ as the amount of network capacity needed to meet both performance guarantees and desired blocking rates at time t, given that the expected number of calls in progress is $q_i(t)$, $i=1,N$. The pricing schedule of guaranteed services should be designed such that $A[q_1(t), q_2(t),...,q_N(t), \overline{\beta}_1(t), \overline{\beta}_2(t),...,\overline{\beta}_N(t)]$ will not exceed total capacity available—that is,

$$A[q_1(t), q_2(t),...,q_N(t), \overline{\beta}_1(t), \overline{\beta}_2(t),...,\overline{\beta}_N(t)] \leq C_T \tag{3}$$

Equation 3 is also called the *admissible region constraint*. By controlling price $p_i(t)$, network operators can control $\lambda_i[p_i(t),t]$, so that the resulting $q_i(t)$ will be kept within the admissible region.

Traffic from guaranteed streams rarely uses up all of network capacity, even when all of the network capacity has been "reserved" for guaranteed traffic. At any given time, the remaining capacity is used for carrying best-effort service. Given that best-effort service does not have strict performance requirement, we will assume that performance is adequate as long as the expected capacity available to best-effort service at time t is sufficient to transmit the expected number of cells admitted into the buffer at time t, for all t—that is,

$$\lambda_b[p_b(t),t] \leq C_T - s_1 q_1(t) - s_2 q_2(t) - ...-s_N q_N(t) \tag{4}$$

OPTIMAL PRICING FOR ATM NETWORK SERVICES

This section discusses the optimal pricing policy for ATM network services. It first presents the optimal pricing model, and then derives and discusses the optimal prices. It demonstrates the advantage of this pricing mechanism over other existing approaches.

The Optimal Pricing Model

From all assumptions and constraints developed in the prior section, the optimal pricing model can be formulated as follows:

$$\text{Maximize} \int_0^T F[p_i(t), \lambda_i(t), p_b(t), \lambda_b(t); i=1,N]dt \tag{5}$$

subject to:

$$\frac{dq_i(t)}{dt} = [1-\overline{\beta}_i(t)]\lambda_i[p_i(t),t] - r_i q_i(t) \qquad i=1,N, \tag{6}$$

and for all t:

$$A[q_1(t), q_2(t),...,q_N(t), \overline{\beta}_1(t), \overline{\beta}_2(t),...,\overline{\beta}_N(t)] \le C_T \tag{7}$$

$$\lambda_b[p_b(t),t] \le C_T - s_1 q_1(t) - s_2 q_2(t) - ...-s_N q_N(t) \tag{8}$$

$$q_i(0)=q_{i0}, q_i(t)\ge 0, \qquad i=1,N \tag{9}$$

where:

$q_i(t)$:   expected number of calls of guaranteed service i in progress at time t,

$p_i(t)$:   expected price for a call of guaranteed service i, starting at time t,

$\lambda_i[p_i(t),t]$: call arrival rate of class i guaranteed service at price p_i, time t,

r_i: call departure rate for guaranteed service i; $^1/_{r_i}$ is average call duration,

$\overline{\beta}_i(t)$:   desired blocking rate for guaranteed service i at time t,

s_i: average transmission rate of guaranteed service i,

$p_b(t)$:   expected spot price for best-effort service at time t,

$\lambda_b[p_b(t),t]$: average rate of admitting cells of best-effort service into buffer at time t,

C_T: total amount of network capacity.

$F[p_i(t), \lambda_i(t), p_b(t), \lambda_b(t); i=1,N]$ is the network operators' utility from providing services at time t. It could be profit or social welfare depending on whether the network is controlled by an unregulated private company or a regulator. The objective of the pricing model is to maximize the service provider's utility function over the time horizon $[0,T]$, as shown by Equation 5. Equation 6 describes how the expected number of calls in progress changes over time; Equation 7 is the admissible region constraint, which guarantees adequate performance to guaranteed services. Equation 8 ensures adequate performance for best-effort service.

Optimal Pricing Schedule

Different pricing policies can be derived from this model based on different assumptions for the utility functions of the network operator. It is generally believed that capacity investment in a telecommunications network exhibits economy of scale. Under this assumption, the provision of ATM network services would likely be a natural monopoly. Therefore, we assume below a profit-maximizing monopolist. We also consider welfare-maximizing prices. However, in the presence of economy of scale, such prices are not likely to recover the cost of capacity investment. Under such circumstances, either network cost must be subsidized from tax revenue—as in the case with subway systems—or we might assume second-best pricing (i.e., welfare maximization subject to a break-even constraint for the natural monopoly provider). This chapter does not consider second-best pricing, but only discusses welfare-maximizing pricing without a break-even constraint.

As shown in Equations 10 and 11, the utility function for a profit-maximizing natural monopolist can be specified as:

$$F(p_i(t), \lambda_i[p_i(t), t], p_b(t), \lambda_b[p_b(t), t]; i = 1, N)$$
$$= \sum_{i=1}^{N} [1 - \bar{\beta}_i(t)] \lambda_i[p_i(t), t] p_i(t) + \lambda_b[p_b(t), t] p_b(t) \qquad (10)$$

and the utility function for a welfare-maximizing regulator is:

$$F(p_i(t), \lambda_i[p_i(t), t], p_b(t), \lambda_b[p_b(t), t]; i = 1, N)$$
$$= \sum_{i=1}^{N} [1 - \bar{\beta}_i(t)] \{ \int_{p_i(t)}^{+\infty} \lambda_i(\xi, t) d\xi + \lambda_i[p_i(t), t] p_i(t) \}$$
$$+ \int_{p_b(t)}^{+\infty} \lambda_b(\xi, t) d\xi + \lambda_b[p_b(t), t] p_b(t) \qquad (11)$$

If one plugs in either of the two objective functions, the pricing model becomes a constrained optimal control problem, which can be solved [7]. The following proposition gives the optimal prices under either profit- or welfare-maximizing objective:

PROPOSITION: Given the pricing model formulated by Equations 5–9:

1. If the network operator's objective is to maximize profit, the optimal prices are:

$$p_i^*(t) = \frac{\varepsilon_i[p_i^*(t),t]}{1+\varepsilon_i[p_i^*(t),t]} h_i(t) \tag{12}$$

$$p_b^*(t) = \frac{\varepsilon_b[p_b^*(t),t]}{1+\varepsilon_b[p_b^*(t),t]} l_2(t) \tag{13}$$

where

$$\varepsilon_i[p_i^*(t),t] = \frac{\partial \lambda_i[p_i^*(t),t]}{\partial p_i^*(t)} \frac{p_i^*}{\lambda_i[p_i^*(t),t]}$$

and

$$\varepsilon_b[p_b^*(t),t] = \frac{\partial \lambda_b[p_b^*(t),t]}{\partial p_b^*(t)} \frac{p_b^*}{\lambda_b[p_b^*(t),t]},$$

and $h_i(t)$ is the Hamiltonian multiplier of the state evolution equation (3–2), $i=1,N$. $l_1(t)$, $l_2(t)$, are Lagrangian multipliers of the constraints described by Equation 7 and Equation 8, respectively.

2. If the network operator's objective is to maximize social welfare, the optimal prices are:

$$p_i^*(t) = h_i(t) \qquad\qquad\qquad i=1,N \tag{14}$$
$$p_b^*(t) = l_i(t) \tag{15}$$

3. $$h_i(t) = \int_t^T [\frac{\partial A}{\partial q_i} l_1(\tau) + s_i l_2(\tau)] e^{-r_i(\tau-t)} d\tau \qquad\qquad i=1,N \tag{16}$$

By comparing the optimal pricing for ATM services in Equations 12 and 13 with the optimal pricing for a tangible product, one can see that the former is quite similar to the latter,[4] except that the marginal cost in tangible product pricing is replaced by other variables in network service pricing. For a guaranteed service, marginal cost is replaced by $h_i(t)$—that is, the Hamiltonian multiplier of Equation 6 (the state evolution equation). For best-effort service, marginal cost is replaced by $l_2(t)$, the Lagrangian multiplier of the capacity constraints (Equation 8). The

---

4. To maximize profit, the optimal price for a tangible product should be $p^*_i = \dfrac{\varepsilon(p^*_i)}{1+\varepsilon(p^*_i)}$ [8].

To maximize social welfare, the optimal price should be $p^* = mc$, where mc is the marginal cost [9].

meaning of these variables can be interpreted as the following: The Lagrangian multiplier of a resource constraint stands for the maximum value that can be derived from one unit of the constrained resource (which, in this case, is network capacity)—or, in other words, the shadow price of consuming one unit of that resource. Equation 8 is the constraint on the amount of network capacity that can be used. Therefore, $l_2(t)$ is the shadow price of using one unit of network capacity at time t, and can be viewed as the opportunity cost of providing one unit of best-effort service at that time.

Similar to $l_2(t)$, $l_1(t)$ can be interpreted as the shadow price for reserving one unit of network capacity. Using $l_1(t)$ and $l_2(t)$ as inputs, one can calculate the opportunity cost of providing type i guaranteed service at time t by Equation 16. At each moment, the opportunity cost of providing type i guaranteed service has two ingredients: (a) the cost of reserving network capacity, which is measured by $l_1(t)$ times $\dfrac{\delta A}{\delta q_i}$, the amount of capacity that needs to be reserved to support a marginal increase in load from class i service, and (b) the cost of using network capacity, which is measured by $l_2(t)$ times s_i, the average transmission rate of the service. The total opportunity cost of a guaranteed service is the sum of opportunity costs at all moments within the service duration. The total cost is obtained by taking mathematical expectation, using the distribution function of the call duration. In this case, call duration is assumed to be exponentially distributed.

Advantages of Proposed Pricing Scheme

It was argued earlier that the opportunity cost of providing a service in ATM networks is determined by the performance objectives, regardless of whether that performance objective is guaranteed, average data rate, burstiness of cell arrivals, call duration, or time of service. Existing pricing schemata are not efficient for ATM networks because they fail to consider differences in all these parameters among multiple services. The following demonstrates that the pricing scheme proposed in this chapter is an improvement to existing ones by showing how all these differences are captured in the pricing schedule.

PERFORMANCE GUARANTEE. In determining the opportunity cost for guaranteed services, this section includes both $l_1(t)$, the shadow price of reserving network capacity, and $l_2(t)$, the shadow price of using the capacity are included (see Equation 16). In calculating the opportunity cost for best-effort service, only $l_2(t)$ is considered. Consequently, guaranteed services will be charged for both reserving and using network capacity, whereas best-effort service will only be charged for using the capacity.

CONSUMPTION OF NETWORK RESOURCE. In this pricing model, best-effort service is charged on a per-cell basis, so price increases with data rate. The price for a guaranteed service is an increasing function of $\frac{\delta A}{\delta q_i}$, which is the amount of capacity needed to be reserved to support a marginal increase in load from class i service, and s_i, which is the average transmission rate of class i service. $\frac{\delta A}{\delta q_i}$ inherently takes into account the average data rate, the burstiness, the blocking probability [10], and the performance objectives. Thus, users who reserve and/or use more network capacity, or who require better performance, will be charged more.

TIME OF SERVICE. It can be seen from Equation 12 to Equation 16 that the optimal price of a guaranteed service is an increasing function of $l_1(t)$, the shadow price of reserving bandwidth, and $l_2(t)$, the shadow price of using bandwidth. The optimal price of best-effort service is also an increasing function of $l_2(t)$. In resource-constrained optimal control problems, the shadow price of a resource increases as the contention for the resource increases. Thus, $l_1(t)$ will increase if there are more requests for reserving network capacity, and $l_2(t)$ will increase if there is more demand for using the network capacity. Consequently, everything else being equal, a service provided during peak hours will be charged more in this pricing scheme than a service provided during a less congested period.

SERVICE DURATION. In the proposed pricing scheme, guaranteed service is charged by a time-dependent usage price, $\phi_i(t)$, for each time unit the call lasts. Therefore, if two users of the same class of service start at the same time, the one who makes a longer call will be charged with a higher price.

As demand changes over time, the shadow prices of reserving and/or using network bandwidth also change. If one type of guaranteed service consistently lasts longer than the other, its price should reflect the contention for bandwidth at times further in the future. This requirement is satisfied in this scheme by incorporating the departure rate, r_i, in the pricing model. As can be seen from Equation 16, r_i can be viewed as a discount rate for shadow prices. If calls last longer, departure rate will be lower and r_i will be smaller. Consequently, the price of a service with a longer expected service duration will depend more on future shadow prices of network capacity than a service with a shorter duration.

In summary, the proposed pricing scheme is more efficient than other existing pricing schemata for ATM networks because it considers important characteristics among different services that other methods fail to capture. Interestingly enough, it can be shown that other pricing approaches mentioned earlier, such as connection time-based pricing, capacity-based pricing, and spot-pricing for the Internet, can be viewed as special applications of the proposed pricing scheme under certain scenarios, as is demonstrated next:

SCENARIO 1. Assume that the network offers homogenous guaranteed services to all users, and the admissible region is linear (i.e. $\frac{\delta A}{\delta q_i}$ = const). By applying the proposed pricing scheme to this scenario, the resulting optimal price per unit of time is:

profit-maximizing pricing:

$$p_i^*(t) = \frac{\varepsilon_i[p_i^*(t),t]}{1+\varepsilon_i[p_i^*(t),t]}\int_t^T[const.*l_1(\tau)+s_il_2(\tau)]e^{-\bar{r}(\tau-t)}d\tau \tag{17}$$

social welfare-maximizing pricing:

$$p_i^*(t) = \int_t^T[const*l_1(\tau)+s_il_2(\tau)]e^{-r_i(\tau-t)}d\tau \tag{18}$$

Scenario 1 is the situation usually encountered in networks that provide only telephone service or only broadcasting-quality video service. Equations 17 and 18 indicate that the optimal prices in this situation only depend on call duration and time of service. This means that this pricing mechanism is reduced to connection time-based price.

SCENARIO 2. Assume that the network only offers guaranteed services, and statistical multiplexing is not supported. Thus, customers cannot share their reserved capacity with others, even if it is not used, and capacity is reserved for each user for all time during [0,T]. In this case $\frac{\delta A}{\delta q_i}$ = c_i, where c_i is the maximum amount of network capacity that a user of service i can use.

By applying the pricing scheme to this scenario, the resulting optimal expected price is:

profit maximizing pricing:

$$p_i^*(t) = \frac{\varepsilon_i[p_i^*(t),t]}{1+\varepsilon_i[p_i^*(t),t]}*c_i*B \tag{19}$$

social welfare-maximizing pricing:

$$p_i^*(t) = c_i * B \qquad (20)$$

where $B = \int_0^T [l_1(\tau) + l_2(\tau)] d\tau$

In this case, the optimal price is a function of c_i, the maximum amount of capacity that a customer can use. In other words, optimal pricing is reduced to capacity-based pricing.

SCENARIO 3. Assume that the network only offers best-effort service, so there is no capacity reservation. If the pricing mechanism is applied to that situation, at each moment, the network will accept cells with highest willingness to pay and charge a spot price that equals the lowest willingness to pay of the cells that enter the buffer at that moment. Thus, the pricing approach is reduced to the spot-price scheme [5].

These three scenarios demonstrate that the proposed approach does not exclude some major pricing schemata that have been used or proposed previously, but rather is a generalization of them. The previous pricing approaches work for single-service networks, in which some service parameters, such as performance objectives and capacity consumption, are the same for all users. As a result, in making pricing decisions, one can take a simpler approach by just ignoring these parameters. However, in ATM networks, there will large differences in all these parameters among multiple services, so a more comprehensive pricing scheme, such as the one developed in this chapter, is needed.

CONCLUSION AND FUTURE WORK

This chapter developed an optimal pricing scheme for ATM networks services. In this scheme, the price for a network service is based on the opportunity cost of providing the service, where the opportunity cost is determined from the shadow prices of using and/or reserving network capacity throughout the expected call duration. The chapter demonstrated that this approach is more efficient than previous network pricing schemata. Important differences in service parameters among different services, which were ignored in previous approaches, are considered in this pricing model.

Future research needs to (a) consider service substitution effect and temporal substitution effect, (b) incorporate these factors into the existing model, and (c) discuss their impact on the optimal pricing schedule. Issues related to the pricing of multilink ATM networks should also be explored.

ACKNOWLEDGMENTS

The authors are grateful to Dr. Bridger Mitchell, Dr. Ingo Vogelsang, and Dr. Benoit Morel for their helpful comments. The remaining errors are ours. Financial support from the National Science Foundation under grants NCR-9210626 and 9307548-NCR is gratefully acknowledged. The views expressed in this chapter, however, are our own.

REFERENCES

[1] Jon M. Peha, *Scheduling and Dropping Algorithms to Support Integrated Services in Packet-Switched Networks*, Ph.D. Dissertation, Technical Report No. CSL-TR-91-489, Computer Systems Laboratory, Stanford University, June 1991.

[2] Ron Cocchi, Scott Shenker, Deborah Estrin, and Lixia Zhang, "Pricing in Computer Networks: Motivation, Formulation, and Example," IEEE/ACM Transactions on Networking; vol. 1, no. 6, December 1993, pp. 614–627.

[3] Jon M. Peha, "The Priority Token Bank: Integrated Scheduling and Admission Control for an Integrated-Service Network," *Proceedings IEEE International Conference on Communications ICC-93*, Geneva, Switzerland, May 1993, pp. 345–351.

[4] Jon M. Peha and Fouad. A. Tobagi, "Cost-Based Scheduling and Dropping Algorithms to Support Integrated Services," *IEEE Trans. Communications,* vol. 44, no. 2, February 1996, pp.192–202.

[5] Hal Varian and Jeff MacKie-Mason, "Pricing the Internet," presented at Public Access to the Internet, JFK School of Government, Harvard University. To appear in Brian Kahin and James Keller, eds., *Public Access to the Internet.* Cambridge, MA: MIT Press, 1995.

[6] Qiong Wang, Jon M. Peha, and Marvin A. Sirbu, " The Design of an Optimal Pricing Scheme for ATM Integrated-Services Networks," to appear in Special Issue on Internet Economics, *Journal of Electronic Publishing*, University of Michigan Press.

[7] M. I. Kamien and N. L. Schwartz, *Dynamic Optimization: The Calculus of Variations and Optimal Control in Economics and Management*, New York: North Holland, 1981.

[8] Jean-Jacques Laffont, *Fundamentals of Public Economics*, Cambridge, MA: MIT Press, 1988.

[9] James L. Pappas and Mark Hirschey, *Managerial Economics*, 5th ed. Chicago, IL: Dryden Press, 1987.

[10] Saurabh Tewari and Jon M. Peha, "Competition Among Telecommunications Carriers That Offer Multiple Services," *23rd Telecommunications Policy Research Conference,* Solomon Island, MD, September 30–October 2, 1995.

6

The Political Economy of Congestion Charges and Settlements in Packet Networks

William H. Lehr
Columbia University

Martin B.H. Weiss
University of Pittsburgh

The dramatic growth of Internet traffic and the expectation that ATM services will play an increasingly important role in the future of the Public Switched Telecommunications Network (PSTN) are attracting new interest in the economics of pricing for packet-based services. Because the costs of these networks are largely fixed, optimal usage prices will differ from zero only to the extent that there are congestion costs. MacKie-Mason and Varian (1992, 1993, 1994, 1995a), Bohn, Braun, Claffy, & Wolf (1993), Clark (1995), Cocchi et al. (1992), Estrin and Zhang (1990), Parris and Ferrari (1992), Parris et al. (1992), Shenker et al. (1995), and others have proposed several approaches for implementing congestion-sensitive pricing in computer networks.

This analysis extends the modeling framework presented by MacKie-Mason and Varian (1995a). This framework is based on a single network domain in which end-to-end network service is supplied by multiple, independent carriers that may neither have the information nor the incentive to cooperate in setting prices or preparing investment strategies that are optimal for the overall network of networks. It is shown here how it may be possible to set optimal congestion prices using only local information on costs and traffic. In addition, this chapter examines the settlements problem that arises with multiple networks, and discusses some of the difficulties this presents for effective implementation of congestion prices.

CONGESTION PRICING FOR INTERCONNECTED NETWORKS

Because most of the costs of constructing and maintaining an electronic communications network, such as the telephone or Internet networks, are largely fixed (or sunk), the carrier's marginal cost for handling additional traffic is close to zero. Therefore, uniform marginal cost pricing will not allow service providers to recover their costs. This has led to wide use of nonlinear pricing strategies, which usually take the form of multipart tariffs that include separate charges for access and usage. When carrier costs are not sensitive to usage, it is possible to recover the bulk of network costs in the form of a flat monthly access fee. As long as the network's quality of service is unaffected by the level of traffic, usage fees may be undesirable. However, if usage is free, consumers will fail to take into account the full social costs of their traffic. These include the reduction in service quality that may be experienced by all subscribers as the network becomes more congested.

Network capacity is limited. As network congestion increases, customers may experience increased delays, higher error rates, or an increased probability that their traffic will be blocked. Although the direct variable costs to the service provider may not be affected, this reduction in service quality may impose large social costs on the aggregate community of subscribers. If it turns out that it is either inexpensive enough or desirable for other reasons to install sufficient excess capacity—so that the network remains uncongested even with zero usage prices (i.e., consumer demand for bandwidth is finite at zero prices)—then these social costs will be small. However, if the network is capacity-constrained, it may be desirable to charge usage prices that reflect the higher social costs associated with increasing congestion.

There are a number of solutions available for allocating scarce bandwidth among competing users. One of the most obvious is "first come, first served." In traditional connection-oriented telephone networks, each customer receives a fixed allocation of bandwidth until capacity is exhausted. Additional calls are blocked. Although simple to implement, this strategy does not discriminate among traffic that may differ widely in its value to customers. This can lead to an inefficient allocation of bandwidth, and can encourage wasteful investments by customers who must compete for the scarce bandwidth. High-value users may be driven to invest in private networks to guarantee access, which could result in higher costs for those who continue to rely on the public network.

A centralized call-admission or traffic-control policy could control this directly, but this would require too much information regarding the exact nature of consumer demands. One obvious alternative is to offer priority pricing: higher prices for higher quality of service and preferential access to bandwidth. This induces consumers to self-sort their traffic in order of value, which can result in significant benefits to both classes of subscribers. Another alternative is peak load or congestion pricing—where users are charged prices that vary with time and the availability of resources. When capacity is scarce, prices should be higher to reflect the increased social costs of congestion. Telephone networks implement a version of this in the form of off-peak discounts for evening and weekend calling.[1]

Specifying the appropriate congestion price makes it possible to decentralize decision making by forcing subscribers to internalize the full social costs (i.e., excess congestion) imposed on all subscribers to the network. It is shown here that, with appropriate assumptions, it may be possible to compute these prices using only knowledge about local demand and capacity cost conditions. Although the rationale for positive congestion prices is derived from the negative impact that congestion may have on all users of the network of networks, it is not usually necessary to know individual responses to increased congestion to set prices. This is important because the individual responses to congestion are not directly observable.

MacKie-Mason and Varian (1995a) provided an analysis of congestion pricing in a single network. Their analysis assumes that all network costs are fixed, and that subscribers benefit when they originate calls but suffer when network congestion increases. Congestion increases with network utilization, measured as the ratio of aggregate traffic to network capacity. Because the only beneficiary of an additional call is the originator, and because each additional call increases network congestion, the social externality is unambiguously negative, which provides the justification for positive congestion prices.[2] Their framework demonstrates that the efficient uniform congestion price is a function of aggregate demand, total capacity costs, and network capacity. It is not necessary to observe individual consumer demands to set optimal congestion prices for an efficiently sized network. Because the individual demands are not readily observable by the service provider, this result is important. Although it is unclear how the carrier selects the efficiently sized network, it is plausible that the carrier might be able to forecast aggregate demand for a single network domain.

This chapter extends the MacKie-Mason and Varian (1995a) analysis to the case of M network domains, which raises several important issues. First, once there are two or more networks, it is no longer clear how one should measure the congestion experienced by a subscriber. In principle, we might expect it to vary depending on the type of calls made (i.e., on-net or internet), the route followed by the call, and the capacities of the various subnetworks.[3] Second, there is the additional problem

---

1. When traffic patterns are relatively predictable, peak load prices, such as those used in telephony, are possible. When the congestion is unpredictable, dynamic prices may be necessary.

2. If the recipients of calls also benefit and this benefit is sufficiently large, then the social externality from additional calls may be positive. Srinagesh (1995) noted that this is one of the rationales for zero settlements among Internet service providers.

3. The term *internet* (uncapitalized) is used to refer to communications across semiautonomous network domains. The Internet is the worldwide TCP/IP packet-switched collection of networks that has evolved from the research-based Department of Defence-funded ARPANET. The Internet is the best known of the many potential Internets to which this analysis may apply.

of settlements, or determining how usage and, potentially, access revenues should be distributed among the multiple carriers. In a dynamically stable, long-run equilibrium, each must recover sufficient revenues to cover its network costs. In general, this will require transferring revenue among the carriers. The mechanism chosen for mediating these transfers (e.g., on the basis of calls handled) may affect carriers' incentives to manipulate their congestion status, which in turn may influence the setting of congestion prices. To address these issues, the earlier modeling framework is modified as follows:

- Let there be M networks, each of which has N_i total subscribers. A type "ij" subscriber makes calls that originate on network "i" and terminate on network "j." These calls are transported across each of the networks along the route followed by type "ij" calls. Let $R(ij) \subset M$ denote the subset of networks that is included in the route of call "ij." To simplify the analysis, it is assumed that each subscriber makes a unique type of call, and that the call follows a unique path through the network of networks.[4] Let $Z=\{ij$ such that $i,j \in M\}$ designate the set of all possible types of calls. The total number of subscribers on the ith network is given by $N_i = \Sigma_{j \in M} N_{ij}$.

- Let $U^{ij} = U^{ij}(x_{ij}, Q^{ij})$ be the utility of a type "ij" consumer, where x_{ij} is the number of type "ij" calls and Q^{ij} is the congestion experienced by type "ij" calls. Following MacKie-Mason and Varian (1995a), assume that utility is weakly increasing in calls originated and is weakly decreasing in the level of congestion (i.e., $\partial U^{ij}/\partial x_{ij} \geq 0$ and $\partial U^{ij}/\partial Q^{ij} \leq 0$).[5]

- The level of congestion, Q^{ij}, provides an inverse proxy for the quality of service experienced by "ij" calls. It could be measured in a wide variety of ways, such as the level of average delay, the maximum potential delay, the bit error rate, the delay jitter, the blocking probability, or some weighted average of all of these. In general, one might expect it to be a weakly increasing function of the volume of each type of traffic and a weakly decreasing function of each network's capacity. The analysis can be further specialized by assuming that congestion is measured in terms of the average end-to-end delay, and that this is simply the sum of the average delay expected at each switching node along the call's route, or

---

4. The assumption that each subscriber makes a single type of call is less restrictive than it may at first appear because a "real-world" subscriber who makes multiple types of calls may be modeled as several different subscribers as long as he or she does not regard different types of calls as close substitutes. This seems reasonable for most types of calling (i.e., a caller in New York does not regard calls to California and Florida as substitutes). The assumption of unique routing may be extended to include connectionless traffic if time intervals are suitably short and R(ij) is allowed to change over time.

5. It is assumed that subscribers ignore the effect their traffic has on overall congestion because N_i and perhaps N_{ij} are large (or, $[\partial U^{ij}/\partial Q^{ij}][\partial Q^{ij}/\partial x_{ij}]$ is close to zero). Note that this does not imply that the aggregate effect on all subscribers of additional congestion is small.

$$Q^{ij} = \sum_{k \in R(ij)} D[Y_k] \qquad (1)$$

where $D[Y_k]$ is the average delay on the kth subnetwork along the route. It is assumed that $D[.]$ is a continuous, monotonically increasing function of network utilization, which is defined as the aggregate traffic handled by network "k" divided by its capacity (i.e., $Y_k = X_k/K_k$).

- The aggregate traffic carried by the ith network, X_i, consists of on-net and internet traffic. On-net traffic both originates and terminates on the same network. The internet traffic may be divided into traffic that originates (terminates) on the ith network, but terminates (originates) on another network and pure transit traffic. The total traffic that originates on network "i" equals $X_i^{On}+X_i^{Off}$, where $X_i^{On}=N_{ii}x_i$ is the on-net traffic and $X_i^{Off} = \sum_{k \in M, i \neq k} N_{ik}x_{ik}$ is the internet traffic. The internet traffic that either terminates on network "i" or is pure transit traffic is given by $X_i^{In} = \sum_{kj \in Z, k \neq i, i \in R(kj)} N_{kj}x_{kj}$. Therefore, $X_i = X_i^{On}+X_i^{Off}+X_i^{In}$.
- Assume two-part tariffs and voluntary participation and that the "sender pays," so that the surplus realized by consumer "ij" is $U^{ij}(x_{ij},Q^{ij})-p_{ij}x_{ij}-T_i \geq 0$ in equilibrium, where p_{ij} is the total congestion charge for call "ij" and T_i is the fixed access charge for network "i."
- Assume that all network costs are fixed and that the costs of each subnetwork depend only on the capacity of that subnetwork. Let the cost of the ith network be described by a continuous, differentiable function $C^i(K_i)$.[6]
- Finally, *social welfare* is defined as the sum of consumer and producer surplus, and there are no external subsidies allowed.

With the previous assumptions and in the absence of settlements, the profit realized by the ith network service provider can be computed as the sum of access and usage revenues less network costs:

$$\Pi^i = N_i T_i + \sum_{j \in M} N_{ij} p_{ij} x_{ij} - C^i(K_i) \qquad (2)$$

The assumption of voluntary participation implies that Π_i must be weakly positive in equilibrium. Total welfare may be computed as:

---

6. In a more general model, one might not expect network costs to be separable, as assumed here. Also, one might expect more complex interactions among different types of traffic and capacity in the determination of call-specific congestion. Furthermore, computing the least cost route for a call may be quite difficult because it amounts to optimally routing traffic so as to minimize congestion costs.

$$W = \sum_{i \in M} \sum_{j \in M} N_{ij} \left(U^{ij} - p_{ij} x_{ij} - T_i \right) + \sum_{i \in M} \Pi^i \tag{3}$$

In the absence of settlements, one finds the optimal congestion prices for an equilibrium-sized network from inspection of the first-order condition for maximizing social welfare with respect to each type of traffic. Each of these first-order conditions is of the form:

$$\frac{\partial W}{\partial x_{ij}} = 0 = N_{ij} \frac{\partial U^{ij}}{\partial x_{ij}} + \sum_{\substack{lk \in Z \\ lk \neq ij}} N_{lk} \frac{\partial U^{lk}}{\partial Q^{lk}} \frac{\partial Q^{lk}}{\partial x_{ij}} \tag{4}$$

The second term is the negative externality imposed on other network subscribers from increased congestion when type "ij" consumers increase their calling. To induce a type "ij" subscriber to internalize the effects of his or her calling, congestion prices should be set so that:

$$p_{ij}^* = -\sum_{\substack{lk \in Z \\ lk \neq ij}} \left(N_{lk} \frac{\partial U^{lk}}{\partial Q^{lk}} \frac{\partial Q^{lk}}{\partial x_{ij}} \right) - (N_{ij} - 1) \frac{\partial U^{ij}}{\partial Q^{ij}} \frac{\partial Q^{ij}}{\partial x_{ij}} \tag{5}$$

The first term on the right side of Equation (5) represents the congestion externality imposed on other subscribers whose traffic is carried on the ith network, whereas the last term is the congestion externality imposed on other type "ij" subscribers. Substituting further for Q^{ij} in Equation (5) and rearranging yields:

$$p_{ij}^* = -\sum_{n \in R(ij)} \frac{D^n_Y}{K_n} \left(\sum_{\substack{lk \in Z \\ n \in R(lk)}} N_{lk} U^{lk}_Q \right) \tag{6}$$

where $D^n_Y = \partial D(Y_n)/\partial Y_n$ and $U^{lk}_Q = \partial U^{lk}/\partial Q^{lk}$. Because network utilization may vary, one cannot assume that the marginal increase in delay is constant for all networks. Therefore, the "n" superscript is retained to remind that D_Y should be computed for each network along the route of call "ij." If one further assumes that network service providers earn zero profits (i.e., that the markets are contestable; see Baumol, Panzar, & Willig, 1982), then one can compute the optimal access charge incorporating the optimal values for X, p, and K into the service providers' profit functions.[7]

---

7. One must check that each consumer's surplus is weakly positive, such that participation is not an issue. It is assumed that this is the case.

With a single network, as in MacKie-Mason and Varian, the optimal congestion price is given by:

$$p^* = -\frac{N-1}{K}\frac{\partial U}{\partial Q}\frac{\partial D}{\partial Y} \tag{7}$$

In the case where M=2, there are only four types of calls: "11" and "22" on-net traffic, and "12" and "21" internet traffic. The formula in Equation 7 can be used to compute the optimal congestion prices for the three types of traffic as follows:

$$p_{11}^* = -\frac{D_Y^1}{K_1}(N_{11}U^{11}_Q + N_{12}U^{12}_Q + N_{21}U^{21}_Q) \tag{8}$$

$$p_{22}^* = -\frac{D_Y^2}{K_2}(N_{22}U^{22}_Q + N_{12}U^{12}_Q + N_{21}U^{21}_Q) \tag{9}$$

$$p_{12}^* = -\frac{D_Y^1}{K_1}(N_{11}U^{11}_Q + N_{12}U^{12}_Q + N_{21}U^{21}_Q)$$
$$-\frac{D_Y^2}{K_2}(N_{22}U^{22}_Q + N_{12}U^{12}_Q + N_{21}U^{21}_Q)$$
$$= p_{21}^* = p_{11}^* + p_{22}^* \tag{10}$$

Thus, the optimal congestion price for internet calls should be equal to the sum of the congestion prices for on-net calls. This is intuitively satisfying because an internet call congests both networks, whereas an on-net call congests only the network that carries it. This result generalizes to the case of M networks: To find the optimal congestion price for a call "ij," one should add the optimal on-net congestion prices for each node along the route (i.e., for the subset of networks in R[ij]).

When the prior pricing results are combined with the first-order conditions used to compute the welfare-maximizing levels of capacity for each of the M networks, the following relationship is obtained:

$$\frac{\partial W}{\partial K_j} = 0 = \sum_{\substack{lk \in Z \\ j \in R(lk)}} (N_{lk} U^{lk}{}_Q \frac{\partial Q^{lk}}{\partial K_j}) - \frac{\partial C^j(K_j)}{\partial K_j}$$

$$= -\frac{X_j D^j{}_Y}{(K_j)^2} (\sum_{\substack{lk \in Z \\ j \in R(lk)}} N_{lk} U^{lk}{}_Q) - \frac{\partial C^j(K_j)}{\partial K_j} \qquad (11)$$

$$p_{ii}^{\bullet} = \frac{\partial C^i(K_i)}{\partial K_i} \frac{K_i}{X_i} \qquad (12)$$

This is analogous to the result in MacKie-Mason and Varian, and shows that it is possible to compute the optimal on-net congestion charge based on local information (i.e., without direct knowledge of the utility functions for the individual subscribers) at equilibrium. As long as each subnetwork charges each packet it carries p_{ii}^*, the total congestion revenues collected by network "i" will provide it with the proper signal for when to expand capacity (i.e., when congestion revenues exceed the value of the subnetwork's capacity valued at the marginal cost of additional capacity).

Three points are worth noting about this result. First, the optimal solution requires that internet traffic should face higher end-to-end congestion charges because it results in more congestion per minute than does on-net traffic. In general, each type of traffic that has a different impact on overall congestion should face a different end-to-end congestion price. This is a form of "congestion priority pricing," which is analogous to other priority pricing schemata in its intent, but is motivated by a slightly different need. In priority pricing, subscribers who are less congestion sensitive accept a lower quality of service in return for a lower price. In the example cited earlier, it would be optimal to charge different rates for internet and on-net traffic even if all consumers had identical preferences with respect to congestion.

Second, the subnetworks will need to account for all of the traffic that passes across their networks to set efficient local congestion prices, and subscribers will have to be billed for the sum of these prices along the least-cost route. One solution is to have a "pay-as-you-go" billing scheme, where each network charges each packet handled its on-net congestion price and bills the consumer directly. Alternatively, the customer could be billed by the originating network, but then the originating network would need to know what the sum of the congestion prices is along the rest of the least-cost route (i.e., $p_{ij}{}^* - p_{ii}{}^*$) to set the appropriate price for a type "ij" call.

If there are at most two networks involved in every internet call (i.e., there are no transit networks), networks could bill each other for terminating calls.[8] This would provide each sub-network with the information about the appropriate termination charge for a call, and the total congestion revenue collected would provide an accurate signal of whether it was advisable to expand capacity.

Another solution is to have the networks continuously update each other regarding their congestion charges, which would allow the originating network to compute $p_{ij}{}^*$ directly. This may be the case in a least-cost routing environment. If routing is hop by hop, the appropriate congestion charge could be passed back up the chain if each node billed traffic the sum of its on-net cost plus the cost charged to terminate the call at the next link in the chain. For example, in a call that will be routed from 1 to 2 to 3, Network 2 should charge network "1" the price $p_{22}{}^* + p_{33}{}^*$, which will allow Network 1 to compute the appropriate end-to-end charge without direct knowledge of Network 3's congestion status.

In all of these solutions, it is possible for the networks to exchange the required information in the form of traffic accounting data without actually making what might amount to sizable revenue transfers in both directions. However, it is important for the networks to account for the congestion charges associated with terminating or transmitting traffic that originates on other networks. Failure to include this traffic may either result in on-net prices that are too high or the failure to invest in adequate network capacity when such investment is appropriate.

Third, although the ability to compute optimal prices based solely on local conditions holds at equilibrium, it is not clear how equilibrium would be attained in a network of networks without the sharing of aggregate demand information among the carriers. Although MacKie-Mason and Varian did not address this point directly, it seems somewhat more plausible in the context of a single network domain that the carrier would be able to forecast aggregate demand. In the network-of-networks context, the individual carrier would need to forecast the demands of all subscribers on all networks to identify the efficient configuration of subnetwork capacities. Although a better understanding of how this equilibrium solution might emerge and its stability properties is obviously important if congestion pricing is to prove useful, further consideration of pricing dynamics is beyond the scope of the present paper. The result presented here is most useful in highlighting the additional complexities introduced when network ownership is fragmented.

---

8. With three networks, the pure transit network could bill the customer and then pay the originating and terminating congestion charges. A version of this occurs in long-distance telephone when the long-distance company pays the originating and terminating local exchange carriers a per-minute access charge.

OPTIMAL CONGESTION PRICES AND SETTLEMENTS

To understand why a settlements problem arises in a network of networks, it is sufficient to consider a simple example with just two networks. Assuming no settlements, optimal congestion prices, and origination-network billing, each network will earn profits of:

$$\Pi^1 = N_1 T_1 + X_1^{on} p_{11}^* + X_1^{off} (p_{11}^* + p_{22}^*) - C^1(K_1) \qquad (13)$$

$$\Pi^2 = N_2 T_2 + X_2^{on} p_{22}^* + X_2^{off} (p_{11}^* + p_{22}^*) - C^2(K_2) \qquad (14)$$

If the network of networks is to recover its costs without external subsidies, the sum of the profits of the constituent networks must be weakly positive. In the absence of settlements, the profits of *each* network must be weakly positive. This imposes a stronger constraint on the optimization problem, and may require distorting the optimal solution to be satisfied.

If the markets were contestable (free entry) or under appropriate rate of return regulation, service providers might be expected to earn zero economic profits. Setting $\Pi^1 = 0$, substituting for the efficient congestion prices and rearranging yields the following result (which is analogous to the result in MacKie-Mason & Varian, 1995a):

$$\frac{T_1 N_1}{C^1(K_1)} = 1 - \frac{\partial C^1(K_1)}{\partial K_1} \frac{K_1}{C^1(K_1)} + \frac{p_{11}^* X_2^{off} - p_{22}^* X_1^{off}}{C^1(K_1)} \qquad (15)$$

The left-hand side gives the share of network costs that must be recovered, via the flat access fees, for the network to recover its costs. The second term on the right drops out if there is only one network, or if traffic flows are balanced and the optimal congestion prices are identical. In either of these special cases, the share of network costs that are recovered via the flat access fee increases toward one as the ratio of marginal to average capacity costs goes to zero. In the multiple network case, however, it is unlikely that traffic flows would be identically balanced or that the optimal congestion prices would be equal.

In the fully symmetric case, with equal numbers of on-net and internet callers and identical costs for each network, the optimal congestion prices, access fees, traffic, and capacity for each network will be identical. There will not be a settlements problem. Consider what happens, however, if the subscribers are distributed asymmetrically, such that a larger share of the Internet callers are located on Network 1. Under these assumptions, the network congestion caused by a call depends on the route followed, but not the direction of the route (i.e., call "12" causes the same congestion as call "21"). Hence, this change should not affect the optimal access and congestion charges faced by consumers.[9] Under the original solution, however, Network 2 will fail to recover its costs.

In the absence of settlements, there are several approaches that may be used to resolve this problem. First, if participation is not an issue, one could allow asymmetric access charges, with Network 2 charging an access fee that is sufficient to recover its higher costs.[10] Although this solution may be efficient, it may not be perceived as equitable. One could argue that it is unfair that consumers on Network 2 face higher access charges because consumers on Network 1 also benefit from the reduction in overall congestion when Network 2's capacity expands.

Second, if we constrain ourselves to uniform access pricing, it may still be possible to implement the efficient capacity and congestion pricing solution by charging higher access fees to all subscribers. In this case, we would need to prevent entry competition for Network 1 because it will earn positive profits at p* and the new, higher T**.

Third, if we constrain ourselves both to free-entry and uniform pricing, then it will generally be optimal to modify both usage and access fees; in general, we will not be able to achieve the same level of total surplus as in the unconstrained problem. This problem arises because, in a zero-profit equilibrium, it is possible to collect sizable congestion revenues from subscribers to induce them to properly internalize the welfare implications of increased calling. These congestion revenues will permit firms to charge lower access fees than would be necessary in the absence of congestion charges. But the sum of these congestion charges and access fees may be insufficient to recover the costs of all of the networks in the optimal solution. In a "sender-keep-all," "no settlements" world, it would be possible for an uncongested, upstream network that originates a disproportionate amount of traffic to collect most of the congestion revenue.

---

9. We are assuming here that the level of network capacity costs depends on traffic patterns and not on the number of subscribers. Although, in general, we might expect network costs to depend both on the number of subscribers and the capacity, K_j (which may depend on the number of subscribers), this need not be the case for several reasons:

• K_j refers to the capacity that is relevant for determining the level of network congestion; this might be the size of the switch, which may depend on X, and not the number of subscribers that generate X;

• capacity may have to be added in fixed increments, and so equal capacity may be optimal for differing numbers of subscribers over a relatively large range; and

• all traffic may be internet traffic, in which case both networks need identical congestion capacity because all calls transit both networks.

10. If consumers could move freely, then we would end up with the fully symmetric case. However, subscribers may not be freely mobile.

IMPLEMENTATION ISSUES

The discussion in the preceding two sections demonstrated that congestion pricing in a network of networks is significantly more challenging than may have been apparent from consideration of the case of a single network domain. The following two sections identify additional complications that need to be addressed before it is practical to implement congestion pricing. Broadly, these can be classified as *technical* and *strategic*. The goal is to suggest important topics for further research, rather than to posit solutions, which is well beyond the scope of the present chapter.

Technical Implementation Considerations

The result that decentralized congestion pricing is optimal is important from a practical perspective. It means that the decentralization of network control, by itself, does not necessitate the sharing of information of the congestion state of neighboring networks. At the optimum, each network can compute a single congestion price based on local demand and cost information. Although this result is encouraging, there are numerous other practical problems that need to be addressed. Estrin and Zhang (1990) considered some of these. A partial list includes: (a) interaction among application types, network architecture, and accounting; (b) type of service considerations; and (c) accounting overhead (i.e., how much will it cost to modify network hardware and software?). These concerns (and others) have given rise to arguments that simple packet counting is not an adequate basis for settlements.[11]

In addition to these issues, there are several additional considerations that require further investigation:

- Congestion prices work by forcing subscribers to internalize the congestion externality caused by their use of the network. If the total congestion price of a packet is the sum of the congestion prices of the networks it traverses, the user must be aware of the congestion price before the packet is sent. This requires that (a) all price information be continuously available to all users (or subnetworks to which users are attached), and (b) the user (or subnetwork) know the route a packet will take in advance. Requirement (a) places an information flow requirement on all of the networks that may be substantial, depending on how the congestion pricing scheme is implemented. Requirement (b) is reasonable for connection-oriented network services, but may not be for connectionless network services, depending on the routing scheme used and the frequency with which congestion prices change.

11. See, for instance, remarks attributed Vinton Cerf (see Cook, 1995). This report also raises the issue of different "business models" of the internet service providers, arguing that MCI's Internet network, as a predominant "transit" network, is currently unprofitable, raising the pressure for some sort of settlements scheme.

- Even if congestion prices are implemented and price information is dispersed appropriately, there is still the question of billing for network service. There have been a number of approaches that have been proposed for accounting and billing in networked information systems.[12] Before any of these approaches can be applied, however, an overall collection and billing strategy must be identified.
- Computing $\partial C^j(K_j)/\partial K_j$ is likely to be difficult in a complex subnetwork consisting of many components. Although the term *capacity* is used fairly loosely here, its precise definition is more elusive, because *capacity* can be affected by network management, congestion control techniques, and so on, in addition to direct investments in network facilities.
- This solution does not easily adapt to multicast.
- It is assumed that any "receiver-pays" scheme will be handled externally, perhaps using technology like NetBill (Sirbu & Tygar, 1995).

The way in which these details are resolved matters. If the originating network supplies the end-to-end price to the user and performs the billing, settlements may be necessary. If each individual network announces price and bills separately, additional user software is necessary (see e.g., Danielsen & Weiss, 1995) to present a consolidated congestion price (and perhaps a bill) to the end user.[13]

The congestion pricing analyzed here does not include multiple service classes, such as "real time" or "best effort." It is widely anticipated by computer science researchers that some form of performance guarantee will be needed to implement real-time traffic (see Ferrari, 1992; Field, 1994). Parris and Ferrari (1992) argued that different service classes require different prices. Stahl and Whinston (1992) considered client–server computing with priority classes. The structure of their analysis can inform the problem of multiple service classes in networks with congestion externalities as well.

Strategic Implementation Considerations

The preceding discussion assumed that network providers do not have market power, and hence will not be able to bias their pricing, network capacity, or interconnection decisions either to extract consumer surplus or protect surplus profits. If market power is significant in a privatized Internet, there will be myriad ways in which service providers may seek to distort either congestion pricing or the settlements mechanism. For example, a transit network that controlled a bottleneck facility would have an incentive to distort its prices for access (interconnection) and usage fees to extract monopoly rents. It may charge lower or higher than op-

---

12. See, for instance, Edell, McKeown, and Varaiya (1995), Mills, Hirsch, and Ruth (1991), Ruth and Mills (1995), or Sirbu and Tygar (1995).

13. This is, in effect, how the telephone network presently works. Users pay a fixed network access fee directly to the local telephone operating company, and receive a separate statement (often in a consolidated bill) from the interexchange carrier (IXC). This bill includes all settlements between the carriers.

timal usage fees, depending on the relationship between inframarginal and marginal subscriber responses.

If monopoly rents are collected by any of the carriers, the settlements mechanism would provide a vehicle for distributing those rents. Bargaining over the distribution of these rents is likely to prove contentious, which will further complicate implementation of a settlements process. Introducing settlements into network profit calculations will influence their behavior. From the discussion in the preceding section, it should be clear that monitoring individual subscriber or subnetwork behavior would be difficult. Hence, carriers may have an incentive to misrepresent their traffic/congestion status to capture a larger share of any settlements revenue. There is a principal–agent problem that must be resolved. Failure to agree on an appropriate settlements mechanism may cause the network of networks to fragment.

In the past, concerns over excess market power provided the justification for regulation of the cable-TV and telephone industries. In recent years, disaffection with traditional regulatory remedies and advances in technology that have reduced entry barriers have encouraged a trend toward increased reliance on market forces. Although the difficulties posed by imperfect competition are worthy of significant research attention, they go beyond the scope of the present chapter. However, even if we restrict ourselves to the (perhaps dubious) case of contestable carrier markets, we cannot presume that all subscribers will be equally represented or influential in determining how future networks will evolve.

For example, in this model, there is a fundamental tension between subscribers who make different types of calls. On-net and internet callers each would like to see the other's traffic minimized, and hence would prefer to see the other face higher prices. This may have implications for customer attitudes toward the efficient implementation of congestion pricing, and toward the debate about emerging notions of "universal service" for the Internet.[14] As noted earlier, efficient prices should discriminate among on-net and internet traffic; nonzero settlements offer one mechanism for implementing these higher prices.

Suppose that the network community can be convinced of the advisability of congestion pricing, and that the debate has turned to the need to discriminate among different types of traffic.[15] Because efficient congestion pricing implies that internet traffic should face higher prices, these callers would have an incentive to argue against price discrimination while on-net subscribers would take the opposite position. Because the chosen settlements mechanism is likely to affect the feasibil-

---

14. There is a sizable community of Internet users that oppose usage-based pricing. Many of these users are concerned about the effects of usage-based pricing on the modes of behavior (such as mailing lists) that they perceive to be valuable. See Love (1995) for an example of this position.

15. This chapter ignores the accounting and implementation costs associated with usage pricing. These may be substantial and when included in the cost–benefit analysis, may make it optimal to employ usage pricing. Assessing the magnitude of these costs is clearly an important area for further research.

ity of implementing price discrimination, there may be a bias from "internet-type" callers in favor of zero-settlements mechanisms.[16] Consider what might happen in negotiations between the subscriber communities of a large and a small network with symmetric calling among pairs of subscribers. In aggregate, subscribers on the larger network are more likely to make on-net calls while subscribers on the smaller network are more likely to make internet calls. Thus, under congestion pricing, subscribers on the larger network should press for a complex settlements mechanism that facilitates charging for termination traffic, while subscribers on the smaller network may argue for zero settlements. The point of this discussion is to suggest how even in the absence of market power by service providers, the political debate over optimal pricing may be distorted by private economic interests.

The failure to adopt optimal congestion prices may influence the choice of where subscribers choose to originate their traffic, although not all subscribers are likely to face the same flexibility. For example, optimal congestion prices should be identical, regardless of the direction in which a particular calling route is followed. However, if $p_{ij} > p_{ji}$, sophisticated callers will have an incentive to originate their calls from Network j. It is not necessary for a caller to physically locate on another network because he or she could use an inexpensive call to set up the return origination call.[17] Generally, rate arbitrage, which results in similar end-to-end congestion charges for traffic with similar congestion (quality-of-service) characteristics, would be welfare improving. However, such arbitrage may not occur on a sufficiently large scale, and may leave unsophisticated subscribers at a disadvantage.

Content providers are another class of sophisticated subscribers who may seek to influence the setting of usage prices (see MacKie-Mason & Varian, 1995b). Generically, one might presume that providers would like to see relatively low network access and usage fees so that consumers have more surplus to spend on content. Ideally, they might like to see network services provided free (subsidized by general tax revenues, which would include nonsubscribers). Alternatively, if the typical content customer is an inframarginal consumer of network services, they may prefer higher than optimal access fees in return for lower than optimal usage fees. Although this scenario need not be the case, it is suggested to illustrate why the establishment of usage pricing is likely to be contentious.

SUMMARY AND CONCLUSIONS

Usage pricing is both desirable and unavoidable for the Internet. There is still much research that needs to be done to better understand both the theoretical and

---

16. This bias may be partially (or wholly) offset if the uniform on-net price or access fees rise in order for the network to recover its total costs.
17. A number of entrepreneurs offered such services to international callers to arbitrage international telephone settlements that resulted in higher prices for calls that originated internationally.

practical issues that arise in a network of networks. This chapter offered a first step toward examining the dual problem of congestion pricing and settlements in such an environment. The chapter extended the analysis of a single network domain, included in MacKie-Mason and Varian (1995a), to the case of multiple networks. This analysis showed that the end-to-end congestion price should equal the sum of the on-net congestion prices of each of the networks along the route. In an efficiently sized network, these prices may be computed using only local cost and traffic information. This is important if network control is to be decentralized.

In the absence of centralized coordination, the networks need to share congestion pricing information so that the originating networks can know what price to set for end-to-end service. A settlements process that requires networks to bill each other for terminating traffic offers one mechanism for conveying this information. This provides one rationale for the linkage between the two problems. A second rationale stems from the need for each network to recover its costs. If prices are set so as to induce optimal consumer behavior, by forcing them to internalize the welfare implications of their behavior for the network of networks, individual firms may fail to recover sufficient revenue in the absence of settlements.

Even in a world where firms do not have market power, revenue transfers among service providers (i.e., settlements) are likely to be necessary. Because the amount of revenue transferred is likely to depend on both the volume of traffic and the price faced by consumers, congestion pricing and settlements issues are not readily separable. This chapter demonstrated this using a simple case of two networks. Further, it argued that the nature of the settlements problem depends on the technology of the networks being used to deliver service, as well as the design of the settlements mechanism.

This analysis focused on the case where carriers do not have market power. If this assumption is not valid, the problem becomes considerably more complex because strategic interactions among the service providers must be considered. In addition, it concentrated on the situation where a network provides a single type of service—as in today's Internet. If multiple service classes exist, as may be necessary with the emerging ATM-based networks, or if a scheme such as the "smart market" (MacKie-Mason & Varian, 1993) or "precedence" (Bohn et al., 1993) is used to provide price-based priority, additional factors may need to be considered. Finally, our analysis is static and does not consider the important question of how the efficient pricing equilibrium is attained, nor whether it is stable. A dynamic analysis raises numerous technical problems that must be solved (not the least of which is the user interface). There are also a host of new economic issues that arise under a dynamic analysis, particularly if a settlements strategy is explicitly included in the analysis. There is clearly much more work that needs to be done in the area of generalizing this analysis from an economic perspective and in applying it to specific network implementations—both statically and dynamically.

ACKNOWLEDGMENTS

This chapter is being published concurrently in the proceedings of the 23rd annual Telecommunications Policy Research Conference, Gerald Brock and Greg Rosston (eds.), by Lawrence Erlbaum Associates, Mahwah, NJ, and in *Telecommunications Policy*. We would like to thank Marjorie Blumenthal, Dave Clark, Jeffrey MacKie-Mason, and Padamanthan Srinagesh for helpful comments and suggestions. Any errors that remain are our own.

REFERENCES

Baumol, W., Panzar, J., and Willig, B. (1982). Contestable markets and the theory of industry structure. Harcourt Brace. New York.

Bohn, R., Braun, H., Claffy, K., & Wolff, S. (1993) *Mitigating the coming Internet crunch: Multiple service levels via precedence*. Technical report, UCSD, San Diego Supercomputer Center and NSF.

Clark, D. (1995, October). *Adding service discrimination to the Internet*. Paper presented at the 23rd annual Telecommunications Research Policy Conference, Solomons Island, MD.

Cocchi, R., Estrin, D., Shenker, S., & Zhang, L. (1992). *Pricing in computer networks: motivation, formulation, and example*. Technical report, University of Southern California.

Cook, G. (1995, September 3). *Summary of the September 1995 COOK report*. Distributed on the telecomreg newsgroup.

Danielsen, K., & Weiss, M. (1995, March). *User control modes and IP allocation*. Technical report, University of Pittsburgh, Pittsburgh, PA.

Edell, R., McKeown, N., & Varaiya, P. (1995). Billing users and pricing for TCP. *IEEE Journal on Selected Areas in Communications*, 13(7), 1162–1175.

Estrin, D., & Zhang, L. (1990). Design considerations for usage accounting and feedback in internetworks. *ACM Computer Communications Review*, 20(5), 56–66.

Ferrari, D. (1992). Real-time communication in an internetwork. *Journal of High Speed Networks*, 1(1), 79–103.

Field, B. (1994). *A network channel abstraction to support application real-time performance guarantees*. Unpublished doctoral dissertation, University of Pittsburgh.

Love, J. (1995). *Future internet pricing*. Available from gopher:// essential.essential.org: 70/0R0-12615--/pub/listserv/tap-info/950310.

MacKie-Mason, J. & Varian, H. (1992, November). *Some economics of the Internet*. Technical report, University of Michigan.

MacKie-Mason, J. & Varian, H. (1993, April). *Pricing the Internet*. Technical report, University of Michigan.

MacKie-Mason, J. & Varian, H. (1994). Economic FAQs about the Internet. *Journal of Economic Perspectives*, 8(3), 75–96.

MacKie-Mason, J. & Varian, H. (1995a). Pricing congestible network resources. *IEEE Journal on Selected Areas in Communications*, 13(7), 1141–1149.

MacKie-Mason, J. & Varian, H. (1995b, October). *Network architecture and content provision: An economic analysis.* Paper presented at the 23rd annual Telecommunications Research Policy Conference, Solomons Island, MD.

Mills, C., Hirsh, D., & Ruth, G. (1991). *Internet accounting: Background.* Technical Report RFC 1272, Network Working Group.

Parris, C., & Ferrari, D. (1992). *A resource based pricing policy for real-time channels in a packet-switching network.* Technical report, International Computer Science Institute, Berkeley, CA.

Parris, C., Keshav, S., & Ferrari, D. (1992). *A framework for the study of pricing in integrated networks.* Technical Report TR-92-016, International Computer Science Institute, Berkeley, CA.

Ruth, G., & Mills, C. (1995). Usage-based cost recovery in internetworks. *Business Communications Review*, 20, 38–42.

Shenker, S., Clark, D., Estrin D., & Herzog, S. (1995, October). *Pricing in computer networks: Reshaping the agenda.* Paper presented at the 23rd annual Telecommunications Research Policy Conference, Solomons Island, MD.

Sirbu, M., & Tygar, J. (1995). *Netbill: An internet commerce system optimized for network delivered services.* Paper presented to Workshop in Internet Economics, Massachusetts Institute of Technology, Boston, MA.

Srinagesh, P. (1995). Internet cost structures and interconnection arrangements. In G. Brock (Ed.), *Toward a competitive telecommunications industry: Selected papers from the 1994 telecommunications policy research conference.* Hillsdale, NJ: Lawrence Erlbaum Associates.

Stahl, D., & Whinston, A. (1992). *An economic approach to client-server computing with priority classes.* Technical report, University of Texas at Austin.

Pricing in Computer Networks:
Reshaping the Research Agenda

Scott Shenker
Xerox PARC

David Clark
MIT

Deborah Estrin
USC/ISI

Shai Herzog
USC/ISI

In a few short years, the Internet has made a dramatic transformation from nerdy enigma to trendy hangout. With its millions of users and diverse application offerings, the Internet is now seen by many pundits as the archetype of the future global information infrastructure. Because of its heavily subsidized origins, commercialization has come late to the Internet. As the Internet confronts this belated and somewhat awkward transition from research test bed to commercial enterprise, there has been much recent discussion about the role of pricing in computer networks. Numerous workshops and conferences have been held on the topic in both the academic community and the network design community. The popular press has also seized the issue as one of broad interest (see, e.g., Markoff, 1993).

In the popular press and network design community, the agenda has been dominated by debates over whether to move from the present system of charges, based on the speed of the access line (so-called "flat pricing"), to basing charges on actual usage. Some contend that usage-based pricing is unnecessary, and would have disastrous consequences for the Internet. Others argue that moving away from flat pricing and toward usage-based pricing is essential for the Internet's efficiency, and therefore is the key to its future economic viability. Unfortunately, little has been clarified by this heated debate except the depth of the participants' convictions. This chapter demonstrates that usage-based charging and flat pricing are really two ends of a single continuum: the difference between them is not one of fundamental principle, but merely of degree, and hybrids of the two approaches are likely to be used in the future.

The academic discussion of pricing in computer networks has concentrated on a rather different issue. This literature typically assumes the necessity of usage-based pricing, and focuses on achieving optimal efficiency maximal welfare in certain simplified models using usage-based pricing schemes. The satisfaction a network user derives from his or her network access depends on the nature of the application being used and the quality of service received from the network (in terms of bandwidth, delay, packet drops, etc.). The network's resources are used most efficiently if they maximize the total user satisfaction of the user community. To achieve optimal efficiency, usage-based charges must equal the marginal cost of usage. Because the physical transmission of packets is essentially free, the marginal usage cost is almost exclusively a congestion cost; congestion costs are the performance penalties that one user's traffic imposes on other users. This optimality paradigm dominates the research agenda. Much of the literature discusses pricing schemes based on computations of these marginal congestion costs.

The main purpose of this chapter is to advocate shifting the research agenda away from the reigning optimality paradigm and toward a more architectural focus. The chapter uses the phrase "pricing architecture" to refer to those components of the pricing scheme that are independent of the particular local pricing decisions and reflect nonlocal concerns, such as how receivers rather than senders can be charged for usage and how to appropriately charge multicast transmissions. These architectural issues, rather than the detailed calculation of marginal congestion costs, should form the core of the research agenda. To motivate this shift in research emphasis, this chapter discusses the economic as well as the mechanistic design issues central to computer network pricing. This treatment of these issues is designed to be accessible to both the network design community and the economic community, with the intention of providing some common context for these two communities, and thereby increasing the opportunity for dialogue.

This chapter has three distinct parts. The first part critiques the optimality paradigm.[1] It is argued that usage charges may, and perhaps should, exceed marginal congestion costs. Moreover, these marginal costs are inherently inaccessible, and so the quixotic pursuit of their precise computation should not dominate the research agenda. The second part presents a rather different paradigm for network

---

1. The authors include themselves in this critique, having adhered to this optimality paradigm in pervious publications (Cocchi, Estrin, Shenker, & Zhang, 1991, 1993).

pricing: *edge pricing*. This term refers to where the charges are assessed, rather than their form (e.g., marginal congestion costs or not). This emphasis reflects the belief that architectural issues are more important than the detailed nature of the charges. The third part describes two fundamental architectural issues and some preliminary design approaches. The chapter concludes with a brief summary. Because much of the discussion requires some familiarity with network mechanisms, Appendix A presents an extremely short overview of the relevant material.

A CRITIQUE OF THE OPTIMALITY PARADIGM

The optimality paradigm may have particular relevance for isolated settings in which the network provider's goal is to maximize welfare, such as in a nonprofit research network or an internal corporate network. This chapter, however, addresses the role of pricing in a commercially competitive environment. The current Internet service provision market has multiple independent service providers (ISPs), and competition appears to be increasing rapidly. This chapter claims that the optimality paradigm is not an adequate foundation for pricing in such a competitive setting.

The optimality paradigm places special focus on marginal congestion costs. This critique is posed in the form of three questions: (a) Are marginal congestion costs relevant? (b) Are marginal congestion costs accessible? (c) Is optimality the only goal?

Are Marginal Congestion Costs Relevant?

It is a standard result that the overall welfare (the sum of provider profit and consumer surplus) is only maximized when prices are set equal to marginal cost, where these marginal costs take all externalities into account. In computer networks, these externalities include congestion effects, where one user's use imposes a performance penalty on other users, and connectivity effects, where a user benefits from other users being connected to the network.

Competition between network service providers typically drives prices to these marginal costs. If the marginal cost prices are sufficient to recover the facility costs of building and operating the network infrastructure, these marginal cost prices are a stable competitive equilibrium,[2] and so computing marginal congestion costs would be central to network pricing schemes. But it is doubtful that such marginal cost prices will recover the full facility costs of computer networks. Within the context of the congestible resource model (MacKie-Mason and Varian, 1995b), marginal costs only cover the cost of the facilities priced at marginal expansion cost

---

2. The term stable means that the revenues cover costs; the term does not refer to any other dynamical properties of the equilibrium.

(i.e., the total congestion costs are equal to the product of the total capacity times the marginal cost of capacity). If the facility cost are a sublinear function of capacity [i.e., $f(x) > f'(x)x$ for $x > 0$], then facility costs will not be fully recovered by marginal cost pricing (see also Strotz, 1965; Mohring and Hartwiz, 1962). Although the cost structure of networks is in flux as technologies rapidly evolve, it seems clear that a large portion of the facility costs arises from the fixed (i.e., not related to capacity) costs of deploying the physical infrastructure. Consequently, in this chapter it is assumed that although marginal congestion costs may be nontrivial, they will be much less than the total facility cost of providing network service.

In such cases, there is no stable competitive equilibrium (see Srinagesh, 1995; Gon and Srinagesh, 1995, for a more thorough discussion of this point). Any stable situation must have some prices that exceed the associated marginal costs. What guides the setting of prices in such a situation? Although there are few general results applicable here, one could argue that the resulting prices of each firm will satisfy the Ramsey condition of maximizing the consumer surplus while still fully recovering costs (because otherwise competing firms would enter and lure customers away by offering more surplus). Thus, in raising prices to increase additional revenue, network service providers will do so in a manner that retains, to the greatest extent possible, the maximality of welfare (because maximizing welfare at a fixed level of profit is equivalent to maximizing consumer surplus at a fixed level of profit).

It is useful, in the following discussion, to artificially break the pricing structure into two distinct pieces.[3] One component of network charges is the attachment fee, which is charged for gaining access to the network; it is independent of any actual or potential usage. The other component is a usage-constraining fee. There are marginal costs associated with both attachment and usage, and welfare is optimized when they are set equal to their respective marginal costs. If all users derived significant benefit from their network connection, the Ramsey pricing scheme would be to raise attachment fees, but keep usage fees at the marginal cost levels, thereby retaining the optimal usage behavior and merely recouping additional revenue from attachment. This is the argument most commonly used to motivate the continued use of marginal congestion pricing in cases where marginal prices by themselves do not fully cover costs.[4]

---

3. Nonlinear pricing policies are ignored here to simplify the discussion.

4. This chapter assumes that network providers are not also controlling, or directly profiting from, the content delivered over their networks. However, in bundled network, such as cable TV, where the application and network transport are sold as a single unit, there are many more opportunities to recover costs. Profits on content and revenue from advertising are important aspects of pricing in bundled networks. Such bundled networks are not considered in this chapter. Rather, it is restricted to the analysis of pricing pure Internet access without bundled services. The nature of the ISP market is still very much in flux, and there may be other sources of revenue in the future, such as renting space on provider-supplied web servers, which may complicate the rather simplified case analyzed here.

Unfortunately, the assumption of a uniformly large benefit from network access does not appear to apply to current computer networks. The low rates of penetration of Internet connectivity and the high rate of churn in subscriptions to online services such as America Online and CompuServe suggest that, in addition to the many users that derive great value from their network connection, there are probably also many other users whose valuation of network connectivity is marginal, and who would disconnect if attachment fees were raised.[5] Thus, it is expected that both usage and attachment prices will affect welfare, and there will be a unique price point that produces a positive optimal welfare. Assuming smoothness throughout, deviating from the optimal pricing point produces welfare changes that are second order in the price deviations. The matrix of second derivatives will depend in detail on the individual utility functions; there is little reason to expect that, in general, consumer surplus is maximized when only attachment fees are raised. See Appendix B for calculations of Ramsey prices in some simple examples.

In addition, the Ramsey pricing scheme could be different for different subpopulations of users. For instance, low-volume users who derive little benefit from being connected to the network would more likely absorb an increase in usage charges without detaching from the network. This is consistent with current practice; some commercial Internet providers charge based on volume to attract low-volume, marginal benefit users who might not otherwise purchase access. In contrast, most large institutions, which typically derive great value from their network connection, pay substantially for the attachment. The traditional and cellular telephony markets also display extensive second-degree price discrimination (i.e., nonlinear pricing schemes where the per-unit price depends on the quantity purchased). There are many different pricing plans, some with lower attachment charges and higher usage charges, and others with the reverse. A similar use of second-degree price discrimination is expected to increase revenue in computer networks.

There are other considerations that suggest that usage charges must remain at significant levels, even if congestion is extremely low (and so marginal congestion costs are extremely low). Assume that an entering network service provider can steal away a subpopulation of users from their current service provider if the entrant can supply this subpopulation with sufficient bandwidth to satisfy their needs at a cost less than the total fee being charged by the current provider.[6] Then one must impose a "core" condition on the pricing structure, mandating that no subset of users can be

---

5. It is possible that, in the grand and glorious future, the global information infrastructure will have a single Internet like network infrastructure and all households will have a single network connection that carries their telephony, television, and data traffic. At that point, it may well be true that essentially all users will have high valuation of their network connection, and so raising attachment fees would be the appropriate way to raise revenue. However, we are a long way from this utopian vision, and therefore should design current network pricing policies to fit the present situation.

6. This assumes seamless interconnection, so switching providers does not affect connectivity. Otherwise, the decision to switch providers involves many other factors besides cost.

charged more than the cost of providing the subset service. If one believes that bandwidth is responsible for any significant portion of the cost of networks, then usage charges must be used to satisfy this core condition. Usage charges are needed to price discriminate between low- and high-volume users. Otherwise, a competing network provider would steal all the low-volume users away by offering a network provisioned at much lower levels with much lower prices. Thus, this core criterion requires that users who regularly consume (or who plan to consume) significantly less bandwidth be charged less, with the difference reflecting the percentage of cost due to bandwidth. Of course, if bandwidth is relatively cheap (i.e., is a very minor portion of the network cost), this "core" argument has little bite.

Are Marginal Congestion Costs Accessible?

As argued above, when prices are required to fully recover costs there is little reason to expect usage prices will equal the marginal congestion costs. Putting that conclusion aside, one can ask the following question: If one attempted to set prices to these marginal congestion costs, could one actually do so? It turns out that computing these congestion costs is quite difficult.

The relationship between what happens to a packet traversing a network and the resulting change in a user's utility is extremely complicated. When one looks at the fate of a single packet, congestion can cause it to be delayed or dropped. Some applications are very sensitive to this extra delay (or being dropped), and others are not. Pricing schemes seeking to achieve optimal efficiency must take these different delay and drop sensitivities into account. Although in simple theoretical models it is convenient to use the abstraction that a user's utility is a function of, say, average bandwidth and delay (e.g., Shenker, 1994), the real world is significantly more complicated (see Clark, 1995a, 1995b, for a discussion of the properties of best-effort traffic). Unfortunately, there is little beyond these simple theoretical models to guide us.

Moreover, most applications involve a sequence of packets; the effect on utility due to the dropping or delay on one individual packet depends on the treatment given to the rest of the packets. For instance, the performance of a file transfer depends on the time the last packet is delivered. For large files, this transfer time depends almost exclusively on the throughput rate, and not on the individual packet delays (see Clark, 1995a, 1995b, for a more thorough discussion of this point). It is extremely difficult, if not impossible, for the network to infer the effect on the transfer time arising from delaying any of the individual packets, especially because the transfer time is also a function of the user's congestion control algorithm. To make matters even worse, often an entire suite of applications is used simultaneously, and then the user's utility depends on the relationship between the delays of the various traffic streams (e.g., a teleconference may involve an audio tool, a video tool, and a shared drawing tool).

The relationship between handling of individual packets and the overall utility is poorly understood, and it changes rapidly with technology (e.g., advances in congestion control could greatly decrease the sensitivity to randomly dropped packets). An important aspect of the problem is that the Internet architecture is based on the network layer not knowing the properties of the applications implemented above it. If one believes that network service providers will sell raw IP (Internet Protocol) connectivity (i.e., they just provide access at the IP level, and do not interpose any application-level gateways), then they have to price based solely on the information available at the IP level. This greatly restricts the extent to which they can adjust prices to fit the particular applications being used (see MacKie-Mason, Shenker, Varian, 1995, for a discussion of the implications of this layering for content provision).

There have been many pricing proposals in the recent literature, and this chapter does not attempt to review them all.[7] The most ambitious pricing proposal for best-effort traffic is the "smart-market" proposal of MacKie-Mason and Varian (1995a); (see also MacKie-Mason, Murphy, Murphy, 1995a, 1995b). In this scheme, each packet carries a "bid" in the packet header; packets are given service at each router if their bids exceed some threshold, and each served packet is charged this threshold price regardless of the packet's bid. This threshold is chosen to be a market clearing price, ensuring the network is fully utilized. The threshold price can be thought of as the highest rejected bid. Having the packets pay this price is akin to having them pay the congestion cost of denying service to the rejected packet. The key to this proposal is incentive compatibility. Users will put their true valuation in the packet because, as in standard second-price auctions, it only affects whether they get service, but not how much they pay. By putting their true valuation of service in the packet header, users will get service if and only if it costs them less than their valuation of the service.[8]

This proposal has stimulated much discussion, and has significantly increased the Internet community's understanding of economic mechanisms in networks. However, there are several problems with this proposal that prevent it from achieving true optimality. First, the most fundamental problem is that submitting a losing bid will typically lead to some unknown amount of delay (because the packet will be retransmitted at a later time), rather than truly not ever receiving service, so the "bid" must reflect how much utility loss this delay would produce, rather than the valuation of service. Thus, accurate bids cannot be submitted without precisely knowing the delay associated with each bid level, and neither the network nor the

---

7. For a few representative examples, see Bradford (1994), Brownlee (1994), Edell, McKeown, and Varaiya (1994), Gupta, Stahl, and Whinston (1994), Jiang and Jordan (1995), Jordan and Jiang (1995), Li and Hui (1994), MacKie-Mason, Murphy, and Murphy (1995a, 1995b), McLean and Sharkey (1993), Murphy and Murphy (1994), Murphy and Murphy (1995), Parris, Keshav, and Ferrari (1992), Parris and Ferrari (1992), Stahl and Whinston (1992), Wang, Sirbu, and Peha (1995).

8. As an aside, note that the pricing scheme is embedded within the architecture in this proposal. The bids are translated into the prices charged.

user knows this delay. Second, there are complications when the packet traverses several hops on its way to its destination. The valuation is an end-to-end quantity (user only cares about the packet reaching its final destination, and does not care about any partial progress), yet the valuation is used on a hop-by-hop manner to determine access at each hop. One would have to extend the bidding mechanism to evaluate the entire path at once, and this entails a distributed multiple good auction of daunting complexity.[9] Third, the bid is on a per-packet basis, yet many applications involve sequences of packets. It is impossible to independently set the valuation of a single packet in a file transfer when the true valuation is for the set of packets.

Wang et al. (1995) proposed a pricing scheme for flows making network reservations (i.e., asking for a quality of service that entails admission control and some assured service level) where prices optimize a given objective function. Gupta et al. (1994) adopted a similar approach for a best-effort network with priorities. As in any conventional economic setting, the optimality of the pricing scheme depends on knowing the demand function. In settings where the supply and delivery are not time-critical, such demand functions can be estimated over long periods of time. However, in computer networks, a user's utility depends on the delay in meeting his or her service request. Hence, one cannot merely consider the long-term average demand, but must also respond to instantaneous fluctuations when setting prices. In addition, as discussed above, rejecting a packet or a reservation lends to some unknown delay rather than an eternal denial of service, making the valuations of the flows not directly related to congestion costs. Consequently, determining optimality in the presence of fluctuating demand is extremely difficult.

The failure of these mechanisms to achieve true optimality is not a failure of imagination, but rather evidence that the task is beyond the scope of any practical algorithm. The keys to efficiency—knowing the service degradation that will result from a particular network action (i.e., how much delay and/or loss), and knowing the user's utility loss as a result of this service degradation—are fundamentally unknowable.

This is not to imply that usage pricing schemes are of little utility. When compared with a situation with no usage-constraining charges, usage charges greatly increase the efficiency of the network. Simulations and calculations in MacKie-Mason et al. (1995a); Gupta, Stahl, and Whinston (1995); and MacKie-Mason and Varian (1995b) clearly demonstrated the significant advantages usage pricing has over free entry. The point here is merely that such pricing schemes do not achieve true optimality, and that the significant efficiency gains demonstrated could probably also be achieved with explicitly suboptimal schemes.

---

9. If one believes that the major source of congestion is at the edge of the network, then one could only apply the smart market at the edge points. This removes the end-to-end versus per-hop problem, and could be used in the edge pricing scheme as the method of charging.

Is Optimality the Only Goal?

The previous section argued that marginal congestion costs are inherently inaccessible. This critique applies equally to attempts to compute Ramsey prices. However, because price deviations away from the optimal points typically produce only second-order deviations in the total welfare, perhaps such deviations are not of much concern. Moreover, in the pursuit of optimality in simplified models, some more basic structural issues have been somewhat neglected. This section identifies some of these structural issues, and urges that they be given significant attention in the design of pricing policies.

Pricing policies should be compatible with the structure of modern networking applications. One of the recent developments in the Internet is the increasingly widespread use of multicast, in which a packet is delivered to a set of receivers, rather than just a single receiver. By sending packets down a distribution tree, and replicating packets only at the tree's branch points, multicast greatly reduces the network's load. Therefore, it is crucial that pricing give the proper incentives to use multicast where appropriate.

Another important aspect of network applications is that the benefit of network usage sometimes lies with the sender of the traffic and sometimes with the receiver(s). Pricing mechanisms should be flexible enough to allow the charges to be assessed to either, or some combination of both, endpoints. This is a very important goal in computer networks; the ability to charge receivers would facilitate the free and unfettered dissemination of information in the Internet, because the providers of such information would not have to pay the cost of transport. Note that this goal is not achieved by the flat pricing approach. Currently, the source's access charge is paid for exclusively by the host institution. This has not yet caused a problem on the Internet, because the elastic and adaptable data applications can easily adjust to overloaded conditions. However, when real-time applications, and other applications that adapt less well to congestion, are in widespread use, the pinch at the source's access point will be felt more acutely.[10]

Pricing policies should also be compatible with the structure of the network service market. There are numerous independent service providers, and many of these are small providers who merely resell connections into bigger provider networks. The interconnection arrangements between providers are somewhat ad hoc (see Srinagesh, 1995; Gon and Srinagesh, 1995) and changing rapidly. Interconnection among these networks is crucial for maximizing social welfare. Pricing

---

10. The ability to assign charges to the receiving end could, in some cases, be handled by a higher level protocol that redistributes the basic charges determined by the network. However, there are several disadvantages to requiring such a higher level protocol: it requires the ability to transfer funds at a higher level, it cannot deal with capacity-based charging; and, in the multicast case, the required information (such as the membership of the group and the network topology) may not be available at the higher layer. Thus, it is preferable to build flexibility of assignment into the basic charging mechanism.

schemes should not hinder interconnection by requiring detailed agreement on pricing policies and complicated per-flow transfers (i.e., a separate transfer for each flow) of money when carrying traffic from another interconnected network. In addition, these independent service providers should be able to make local decisions about the appropriate pricing policies. This implies that the pricing policy should not be embedded into the network architecture. Instead, the network architecture should provide a flexible accounting infrastructure that can support a wide variety of locally implemented pricing schemes. For instance, there are some contexts (such as managing an internal corporate or university network) where the goal of pricing is merely to encourage efficient use of the network resources. Often in these contexts there are incentives that can be used (e.g., quotas) instead of money. Although this chapter has focused on monetary incentives, the underlying accounting structure and pricing architecture should allow for the use of these other incentive forms if they are locally applicable.

Achieving optimality necessarily involves uniform implementation of a single pricing scheme across the network. Optimality involves setting prices at exactly the marginal congestion costs, thus the accounting scheme becomes a distributed computation of those congestion costs. Thus, the optimality paradigm is fundamentally inconsistent with the need for locality in pricing. Given that no pricing scheme has claim to being truly optimal, the need for local control should take precedence over the desire for absolute optimality.

Although true optimality is not an appropriate goal, pricing should still be used to achieve reasonable levels of efficiency. It is important that the underlying accounting infrastructure allow prices to be based on some approximation of congestion costs. There is an important distinction lurking here. It is important to allow prices to be based on some approximation of congestion costs, but it is important to not force them to be equal to these approximate congestion costs. As argued, earlier, the need for full cost recovery militates against such an assumption of equality. Meeting any reasonable efficiency goal, however, would likely require that prices depend on such congestion costs.

Rather than start with mechanisms designed to precisely calculate marginal congestion costs, one might first ask: What are the absolutely minimal requirements for providing some estimate of congestion costs? One minimal requirement is that pricing should encourage the appropriate use of quality of service (QoS) signals (i.e., the signals sent by applications to the network requesting a particular quality of service; see Appendix A). This is crucial for making the new QoS-rich network designs effective, and would enable them to achieve significant increases in network efficiency. An additional requirement is that pricing should discourage network usage during times of congestion, but not discourage it during relatively uncongested times. The basic point is that perhaps these minimal requirements are sufficient to achieve reasonable approximations, and that attempts to more accurately calculate Ramsey prices are of little (indeed second-order) value, distracting one from the more important, but often overlooked, structural concerns.

A NEW PRICING PARADIGM: EDGE PRICING

After having critiqued the reigning optimality paradigm, a different pricing paradigm is presented: edge pricing. The edge pricing paradigm is described using a series of approximations to true congestion costs.

Approximating Congestion Costs

Computing the true congestion costs requires that one can compute other users' loss in utility due to one user's use. This requires knowledge of the utility of users, which in the Internet architecture is fundamentally unknowable, as well as knowledge of the current congestion conditions along the entire path. Such detailed knowledge entails a sophisticated accounting scheme that transcends administrative boundaries by following the entire path. Having already concluded that the estimates of utility loss are extremely rough, can one also replace the knowledge of current congestion conditions along the entire path with a reasonable, but more easily accessible, estimate? Consider the following two approximations.

The first approximation is to replace the current congestion conditions by the expected congestion conditions. This is essentially QoS-sensitive time-of-day pricing. The time-of-day dependence builds in expectations about the current congestion conditions. The QoS dependence reflects that the effect one flow's packets have on another flow's packets depends on the respective service classes of the flows. Packets in higher quality service classes impose more delay on other packets than do packets in lower quality service classes. This approximation of QoS-sensitive time-of-day pricing is problematic because it does not reflect any instantaneous fluctuations in traffic levels. Packets sent during a lull in the network would still be charged full price even though the actual congestion costs were quite small. Such insensitivity to instantaneous conditions would seem to remove any incentive for users to redistribute their load dynamically. Just as in the telephone network, time-of-day pricing encourages users to time shift their calls to later (or earlier) hours when rates are lower, but does not encourage them to adjust to the instantaneous conditions. (Of course, in the telephone network, there is no way for users to detect the current load.)

The inability to charge less during periods of low congestion is not a serious problem because, in many cases, one can substitute the congestion-sensitivity of service for the congestion-sensitivity of prices. During a lull in the network, lower quality classes give as good a service as high-quality classes do during congested periods. Users who monitor the service they are getting from the network and adjust their service request accordingly can take advantage of this variability. The way user costs are lowered during times of reduced network load is not that the network lowers the price of service classes, but that users request lower quality service classes and are charged the lower price of that class.

"Users" are referred to as the entities adapting to current conditions to distinguish this from the network adapting. In reality, adaptation does not require significant effort from the human user (see MacKie-Mason et al., 1995b, for a similar discussion of the role of adaptation). Instead, adaptation routines will be highly automated and embedded within applications or the end system's operating system. Many current network applications are already designed to adapt to network conditions. Hence, relying on users to adapt to current condition, rather than the network, is quite consistent with current practice. In fact, this reflects a basic Internet design philosophy. To the extent possible (and routing is the once place where it is frequently less possible), the intelligence and responsibility to adapt to current conditions should be placed on the outside of the network; the fundamental infrastructure inside the network should remain fairly simple, intentionally ignorant of the applications it is supporting, and should not try to adapt on behalf of these applications. Applied to this case, this philosophy argues for relatively static pricing policies, with end users varying their service requests in response to current congestion conditions. This removes from the network the responsibility of accurately assessing current conditions and their likely impact on users' utilities, and puts the onus on individual applications/users to make that assessment for themselves. Given that applications have different sensitivities to service quality, it seems preferable to place the onus of control where it can be done in the most informed way.

If expected congestion were the only approximation, one would essentially have a pricing scheme where prices were computed per link based on the time of day and quality of service requested. The second approximation is to replace the cost of the actual path with the cost of the expected path, where the charge depends only on the source and destination (s) of the flow, and not on the particular route taken by the flow. From a user's perspective, they have requested service from one point to another (at least in the unicast case); the actual path the data take is typically determined by the network routing algorithms (except in the case of source or explicit routing). Having the price of the service depend on the network's decision about routing seems an unnecessary source of price variation that makes it harder for the user to make informed plans about network use. Moreover, when alternate paths are taken by the network in response to congestion, the extra cost due to the congestion should not necessarily fall only on those flows that have been redirected. Certainly in the telephone network, the price of the telephone call does not depend on the network's choice of route.

Edge Pricing

When one combines these two approximations, the price is based on the expected congestion along the expected path appropriate for the packet's source and destination. Therefore, the resulting prices can determined and charges locally assessed at the access point (i.e., the edge of the provider's network where the user's packets enter), rather than computed in a distributed fashion along the entire path. This local scheme is termed edge pricing. A similar approach to pricing in computer networks was suggested by Jacobson (personal communication). The prices charged at the edge, or access, point may depend on information obtained from other parts of the network, but the entire computation of charges is performed at the access point. A later section discusses the multicast case where the relevant information is difficult to obtain.

As discussed by Clark (1995a, 1995b), edge pricing has the attractive property that all pricing is done locally. Interconnection here involves the network providers purchasing service from each other in the same manner that regular users purchase service. When a user connected to Provider A's network sends a packet, it is applied to that user's bill according to whatever pricing policy Provider A has.[11] If the destination of the packet is on Provider B's network, then when the packet enters Provider B's network the packet is charged against Provider's A bill with Provider B. There are no per-flow settlement payments, i.e., the various providers do not redistribute the charge levied to the end user among themselves. Instead, each provider takes full responsibility for every packet he or she forwards; a sequence of bilateral agreements between the adjacent service providers along the path performs the necessary function of cost shifting. These bilateral agreements apply only to the aggregate usage by these providers, and so greatly simplify the transfer of payments between providers.

The beauty of this is that billing structures are completely local. The exact nature of the pricing scheme is simply a matter between the user and service provider. Because the decisions are local, service providers can invent ever more attractive (and complicated) pricing schemes, and can respond to user requirements in a completely flexible fashion. No uniform pricing standards need be developed because interconnection involves only bilateral agreements that allow each provider to use his or her own pricing policy. Locality allows providers to experiment with new pricing policies and gradually evolve them over time. In fact, pricing policies will likely be one of the important competitive advantages available to providers when competing with each other. For instance, locality allows providers to offer specialized pricing deals, such as bulk discounts. It is hard to imagine implementing a meaningful bulk discount when charging is done in a nonlocal, per-link basis. A user's usage of any particular link, or of any particular service provider outside of the local one, is probably quite limited, and so such discounts are much less meaningful.

---

11. The term bill is used here only to connote that the packet is applied to the contract the user has with Provider A. As mentioned later, the contracted pricing policy may be a flat price with a limit on peak rate, in which case there is no additional charge per packet.

Forms of Pricing

Edge pricing describes the place at which charges are assessed, but is completely neutral about the nature of these charges. In most of the literature, there is a sharp distinction between usage and attachment charges; this differentiates the fixed (or flat) portion of the price and the variable usage-dependent portion of the price. Thus, the cost of upgrading the speed of the user's access line is considered an attachment charge. This division is somewhat misleading because there is a natural continuum between the two.[12] Thus, this chapter refers to them all as *usage-constraining prices*.[13] Per-packet charges are clearly designed to constrain usage, but so are limits on a user's peak sending rate.

The continuum of usage-constraining charges can perhaps best be explored by defining its two endpoints. At one end of the continuum, prices can be based on actual usage in the form of per-packet and/or per-reservation charges. This is the traditional form of usage-based pricing. At the other end of the spectrum, users could purchase a capacity from the network and then be allowed to use up to that capacity without any additional charge. One form of capacity could be defined in terms of a peak rate, as in the current form of flat pricing. More generally, however, this capacity is defined in terms of a filter that is applied to the traffic. A usage filter characterizes flows as either conforming or not conforming to the agreed capacity. Such filters can measure the usage over differing time horizons, such as controlling the long-term average rate, the short-term peak rate, and intermediate burst durations. This capacity framework is merely a generalized version of current flat-rate pricing schemes; the extra flexibility allows pricing schemes to be more closely attuned to user requirements (see Appendix C for a more complete explanation of such filters). Although not essential to this discussion, there can be several possible actions that the service provider could take when a user exceeds her capacity. For instance, all such packets could be mapped into the lowest service class, dropped, queued until the flow is in compliance with the filter, or merely assessed an additional per-packet fee.

The units of usage that are applied against the capacity constraints, just like per-packet charges, can depend on many things, such as time of day, destination, and QoS. High-quality service classes might consume twice as many units as lower qual-

---

12. The distinction between fixed and variable prices may be extremely important to individual users. Users on fixed budgets may need fixed prices, whereas users with extremely variable demands may need the ability to only pay for usage. The point is that this distinction, although important to individual users, is not fundamentally important from an architectural or economic perspective. Both forms of pricing can be assessed locally, and both constrain usage.

13. In this taxonomy, attachment prices would refer only to the price of attaching to the network, and not refer at all to the speed of the access line. All other charges would be considered usage-constraining. Of course, in nonlinear pricing schemes (or when there is a spectrum of pricing menus offered, as in the current cellular telephony market), the distinction between the two is completely blurred.

ity service classes, with similar increments for packets traveling further or over particularly congested links. Of course, to realize the goal of allowing users to send an unlimited amount of traffic when the network is empty, there should be a category of absolutely lowest quality of service that is essentially free. In fact, one could even use a "smart market" auction approach to pricing at the access point.[14]

These capacity constraints allow network providers to make informed provisioning decisions. Of course, provisioning decisions will also be heavily based on measurements of actual aggregate usage, but the capacity filter parameters give some additional input for estimates. If there is an infinite amount of multiplexing (i.e., each user constitutes an infinitesimal share of the aggregate usage) and users are uncorrelated, provisioning need only be based on the long-term average rates. The other capacity filter parameters (such as typical burst size) are needed to make estimates of the magnitude of usage fluctuations away from this average value.

Because the overlimit behavior (i.e., when usage exceeds the capacity) can merely be an additional per-packet charge, there can be a continuum of pricing policies that stretch between purely usage-based and capacity-based charging. Within the spectrum of edge charging, the difference between capacity-based and usage-based prices is not a fundamental architectural issue. It is expected that the market will invent, over time, increasingly attractive and flexible hybrids of these approaches. Telephony may provide an instructive example. Telephone companies offer a menu of local calling plans—some usage-based (e.g., metered service), some capacity-based (e.g., unlimited service), and some a combination of both (e.g., a certain number of free minutes per month, plus a metered rate for calls in excess of this number). It is likely that the same will happen in computer networks, with some users choosing usage-based and others choosing capacity-based charges, and many being somewhere in between. Thus, the heated debate between advocates of usage-based and capacity-based pricing schemes will become completely irrelevant as users vote with their feet. In the edge pricing paradigm, the decision between usage-based and capacity-based, or anywhere in between, is completely local, and network provision is competitive. Hence, the offered plans will likely reflect the true needs of consumers (and thus the architecture need not preclude one choice or the other to prevent providers from exploiting users). The rest of this chapter is devoted to exploring the infrastructure needed to support this edge pricing approach.

ARCHITECTURAL ISSUES

Edge pricing localizes the whole charging process—everything occurs at the access point. Yet there are two inherently nonlocal aspects of pricing: (a) charging appro-

---

14. There may be some disadvantages with using the smart-market as the local pricing scheme (e.g., it embeds the pricing policy in the architecture). However, the point here is that it is not architecturally precluded by the edge pricing paradigm, and so firms are free to experiment with it.

priately for multicast, and (b) the ability to charge receivers for the service.[15] These nonlocal aspects pose some fundamental architectural challenges to the edge pricing approach. These issues are viewed as forming the basis of a fertile research agenda in pricing in computer networks. This section discusses how the infrastructure might be designed to handle these nonlocal aspects. It describes the problems of multicast and charging receivers separately, and then reviews some remaining open problems.

It is important to note that this design discussion is extremely preliminary, and is intended to be illustrative rather than definitive. That is, the purpose here is to illustrate some of the issues involve by engaging in a design discussion. However, the design directions advocated here may not, in the end, be the appropriate choices.[16]

Multicast

When unicast packets enter a provider's access point, the destination field is enough to determine the typical path of the packet. Unicast routes fluctuate occasionally, but the normal case is that unicast routes change on rather slow time scales. Thus, fairly static tables at the entry points can provide adequate information for pricing decisions, and it would be relatively trivial to design the distributed algorithms needed to construct and maintain these tables.[17] If addresses encode geographic information (as in Deering's recent proposal; S. Deering, personal communication, 1994) or provider information (as in the current IPv6 proposal; Deering and Hinden, 1995), then these tables are especially simple (see Francis, 1994, for more information). Moreover, if the provider networks are small enough, one set fee for all intraprovider packets and another fixed fee for all interprovider packets might be sufficient.

---

15. Charging receivers for service is a nonlocal problem because, in approaches with explicit willingness-to-pay signaling, when both the source and receiver are serviced by the same provider, the source's access point must be informed that the receiver is willing to assume responsibility for the transmission. Similarly, when the path from source to receive traverses several different provider networks, the notification of receiver paying must be communicated to both the exit and entrance access points in each network. Other approaches can avoid this explicit nonlocal signaling by adopting some uniform standards, such as a certain portion of the multicast address space being set aside for receive-pay groups. But these standards are nonlocal in that they represent agreements between providers about a billing policy.

16. In particular, there is a spectrum of design choices providing different levels of functionality and requiring more of less additional mechanism. The following discussion is not attempting to make a detailed evaluation of the functionality versus mechanism trade-off, but is merely illustrating some possible ways to achieve the aforementioned goals. There are more minimalist approaches to these problems that require less additional mechanism, and they should be considered when making design decisions for the Internet. For this pedagogical discussion, more straightforward, if more mechanistic, approaches have been presented.

17. These tables would contain information describing how many usage units (for a capacity filter) each packet represents, or a monetary per-packet charge, or whatever other information is needed for the provider to assess the appropriate charges.

Multicast packets pose more of a challenge. A multicast address is merely a logical name; by itself, it conveys no geographic or provider information. Although multicast routing identifies the next hop along the path for packets arriving at an interface, multicast routing does not identify the rest of the tree. Thus, estimating costs in the multicast case requires an additional piece of accounting infrastructure. Moreover, the set of receivers—the members of the multicast group—can change quite rapidly and so the mechanisms for providing the appropriate accounting information must be designed with care.

One can imagine several different approaches. The simplest would be to merely collect the location (i.e., subnet numbers) of all receivers (with receivers outside of the provider's network being recorded as residing at the appropriate exit point of the network). From these locations, one could compute the approximate costs of the appropriate tree.

Another approach would be to compute these costs on the fly by introducing a new form of control message—an accounting message—that would be initiated when the receiver sent its multicast join message (*multicast join messages* are the control messages sent by a receiver to join the multicast group; see Appendix A). These accounting messages would be forwarded along the reverse trees toward each source, recording the "cost" of each link it traversed and summing costs when branches merged. When these accounting messages reached a source's access point, the cumulative cost of reaching all receivers from that source would be available. Each provider would only need to record the cost information local to his or her network, (i.e., the costs would start accumulating when the accounting message entered the provider's network, and would stop when the accounting message exited the network). No cost information crosses the provider boundaries. Instead, this cost information is only used locally to compute the charges to apply on the edge of the network. This on-the-fly approach makes the charge for multicast depend on the true path, rather than the typical path, which may cause unnecessary variability.[18] There might be groups (e.g., cable-TV channels) where the typical tree might be well enough known in advance so that such additional mechanisms are not needed.

The additional piece of accounting infrastructure needed to compute these costs is local to the provider (i.e., each provider can use its own algorithm). No standards need be established, and no agreements with other providers need be made. Thus, this protocol can incrementally evolve over time as the cost structure and traffic patterns of future networks are better understood. Independent evolvability is one of the biggest advantages of the edge pricing paradigm. Although the total amount of mechanism needed to perform the necessary accounting may not be

---

18. One could apply such a scheme to a logically overlaid network so the prices would be less dependent on the details of the path. For instance, the network could be divided up into area codes, with local link costs recorded whenever the accounting message left one area code and entered another.

less than in other paradigms, the degree of independence of these accounting mechanisms is substantially higher. The ability for providers to act independently to upgrade their accounting will lead to rapid development of the required implementation. Proposals that require a single uniform and standardized accounting infrastructure are much less likely to ever be implemented.

The prior discussion applies to the complete spectrum of usage-constraining pricing schemes from usage-based to capacity-based charging. However, much of the discussion implicitly applied to best-effort service. The basic principles remain the same when pricing for reserved or ensured levels of service, but the mechanistic details are quite different because of the presence of the setup protocol like RSVP (Zhang, Deering, Estrin, Shenker, & Zappala, 1993).

Charging Receivers

The second nonlocal problem considered here is assigning charges to receivers. This involves addressing the following three issues:

1.   How does a receiver indicate to the network provider that it is willing to take responsibility for the source's traffic? There are several alternatives. A few are mentioned here to illustrate some of the possibilities. In the best-effort multicast case, the join message might be extended to include a willingness-to-pay field. In the case of reservations, either unicast or multicast, the RSVP reservation message could carry similar information. The only case that does not already have a preexisting control message that could be used for this purpose is the unicast best-effort case. Here, one may require a new willingness-to-pay control message to be generated by the receiver, but there may also be other approaches. In addition, one may want to allow the source to indicate that it is not willing to pay, so that if the source's access point has not received a notification that the receiver(s) is (are) willing to pay, the packets are immediately dropped. Such an indication that the source is not willing to pay could be contained in the packet header. Another approach, one that requires no additional signaling, is to divide the multicast addresses into sender-pays and receiver-pays categories, so that the assignment option is indicated by the choice of multicast address. Here the very act of joining the group communicates a willingness to pay.

2.   How does the network "bill" the receiver? One general approach here is to apply pricing when the packets traverse the receiver's access point. Thus, packets are "charged" according to the receiver's contract with its provider, not according to the sender's contract. If the capacity is exceeded, the overlimit behavior (delaying, dropping, etc.) is applied at this exit point. If the packet traverses several providers, this reverse charging is applied whenever a boundary is crossed. The packet is charged to the provider whose network the packet is entering, not the provider whose network the packet is exiting.

3.   How does the network split the responsibility for the bill among the members of a multicast group? If there are multiple receivers, the network needs to transfer the charges to the receivers, as well as apportion the cost among them in a reasonable manner. One way to do this is to assign fractional responsibilities to each of the receivers. Then, when the packet arrives at each receiver's access point, the receiver is "charged" only the fraction of the normal amount. The variety of policies for assigning these fractions, as well as mechanisms for computing them, were addressed in Herzog, Shenker, and Estrin (1995). One could also use cruder approximations to compute these fractions, basing multicast prices on ad hoc discounts from the unicast cost.

Open Issues

The preceding general discussion merely presented some possible approaches. There are many other possibilities; the ones mentioned should be considered sketchy illustrations of the issues involved, rather than serious and complete design proposals. This initial design discussion leaves several fundamental issues unresolved; a few of these are mentioned here.

The discussion of the need for charging receivers has focused on a narrow binary choice: either sources pay or receivers pay. One may want to consider a much broader spectrum of policies in which the costs are shared in a more flexible manner. This might be a fractional splitting (e.g., the source pays 30% and the receiver pays 70%, or perhaps the source pays for a certain portion of the path (e.g., the source pays for the portion of the path within its local provider's network) and the receiver pays the rest. The requirements of such source receiver cost sharing have not been addressed.

In addition, the case where some receivers are willing to pay and others are not has not been considered. Aside from the mechanistic questions, there are important unresolved policy questions about how to handle such a situation. Related to this is that receivers may want to limit their exposure. The willingness-to-pay field may, in addition to indicating that the receiver is willing to pay, also indicate a cap on how much (in some arbitrary units) cost the receiver is willing to absorb. Such a cap may be necessary when joining what one expects to be large multicast groups. For instance, when a receiver in California joins a group for a virtual rock concert sourced from London, with an expected audience of millions, the receiver may be willing to pay his or her share of a few dollars (or equivalent capacity units), but would certainly not be willing to absorb the bill for the entire 50 mbps video feed from London. However, such limits open up thorny strategic issues because receivers would be tempted to free ride on other receivers.

There may be other approaches to deal with the startup phase of multicast groups that will eventually become large. There may be some way that the organizer of the session whether it be a rock concert, an IETF broadcast, or a cable-TV channel could describe to the network beforehand an approximation to the likely distribution tree. This would enable the network to estimate the likely cost shares beforehand, and thus greatly reduce the exposure of the first few group members. Other schemes to reduce such exposure have been discussed in Gawlick (1995).

The accounting mechanisms discussed in Herzog et al. (1995), which determine the appropriate multicast cost shares, are implemented on a link-by-link basis. Such methods must be extended to a more abstract set of logical links so that the cost shares can reflect a coarser level of granularity. Also, as discussed in Herzog et al. (1995), some cost-sharing approaches depend on the number of receivers downstream of each link. Such numbers are relatively easy to obtain within a provider's network (e.g., by extending the multicast join mechanism). To do this accurately across providers would require each provider to reveal this number to other providers. This raises an incentive question because the cost share increases with the number of receivers (and so each provider would reveal only the existence of one receiver in its network).

Another issue arises in the case of receiver-pays with capacity-based charging, where the overlimit behavior is packet dropping. If the incoming traffic greatly exceeds the available capacity, the network has transported packets across the network only to consistently drop them at the exiting access point. Some slow, out-of-band signaling may be needed to unjoin the receiver from the group. Such signaling is not needed if the overlimit behavior is an additional per-packet charge.

CONCLUSION

Current discussions about pricing in computer networks are dominated by two main topics. The first topic is the debate between usage-based pricing and flat pricing, which has embroiled the network design community and caught the attention of the popular press. Rather than being radically different, these two schemes reside along the single continuum of usage-constraining pricing policies. As in telephony, both pricing options, along with various intermediate hybrids, will likely be offered to users by their local provider. The detailed design of such schemes is perhaps best left to the marketing departments of the various network service providers. Thus, no particular pricing policy should be embedded into the network architecture. The challenge for the network design community is to provide a coherent network pricing architecture that allows individual providers to make their own choices about how to price service. This chapter presented one such pricing architecture that achieves this goal: edge pricing.

The second topic, emphasized in the more academic literature, is the design of marginal cost-pricing schemes that produce the optimally efficient use of network resources. This optimality paradigm has been critiqued on three grounds: (a) marginal cost prices may not produce sufficient revenues to fully recover costs, and so are perhaps of limited relevance; (b) congestion costs are inherently inaccessible to the network, and so cannot reliably form the basis for pricing; and (c) there are other, more structural, goals besides optimality, and some of these goals are incompatible with the globally uniformity required for optimal pricing schemes. For these reasons, it is contended here that the research agenda on pricing in computer network should shift away from the optimality paradigm and focus more on structural and architectural issues. Such issues include allowing local control of pricing policies, fostering interconnection handling multicast appropriately, and allowing receivers to pay for transmission. To illustrate this point, this chapter described how these goals might be accomplished in the context of the edge pricing paradigm.

Although many of these detailed comments concern the particular pricing architecture of edge pricing, the intent in writing this chapter was not to advocate that this scheme be adopted by the Internet. The proposal is extremely preliminary, and there may be other schemes with similar properties. Rather, the intent was to initiate a dialogue about such pricing schemes and stimulate the creation of other pricing paradigms that meet these design goals.

ACKNOWLEDGMENTS

This research was supported, in part, by the Advanced Research Projects Agency, monitored by Fort Huachuca under contracts DABT63-94-C-0073 (SS), DABT63-94-0072 (DC), and DABT63-91-C-0001 (DE,SH). The views expressed here do not reflect the position or policy of the U.S. government. This chapter is also being published in *Telecommunications Policy*. We are grateful to Jeffrey MacKie-Mason, Hal Varian, Marjory Blumenthal, Padmanabhan Srinagesh, and Steve Deering for insightful comments on an earlier draft. We would also like to thank Van Jacobson for early conversations on edge pricing.

APPENDIX A: INTERNET ARCHITECTURE AND MECHANISMS

This appendix describes a few relevant features of the Internet architecture. It is a selective and sketchy overview, intended merely to provide a minimal background for reading this chapter. The current Internet architecture is designed for point-to-point (or unicast) best-effort communication. Every packet header contains a source address and a destination address. Upon receiving a packet, a switch (or, equivalently, a router) consults its routing table to find the appropriate outgoing link for the packet, based on the packet's destination address. The network makes no commitments about when, or even if, packets will be delivered. Sometimes the incoming rate of packets at a switch is greater than the outgoing rate, and so queues build in the switch. These queues cause packet delays and, if the switch runs out of buffer space, packet discards. The network does not attempt to schedule use. Sources can send packets at any time, and the network switches merely exert their best effort to handle the load.

This simple network architecture has been amazingly successful. However, there are efforts currently underway to extend this architecture in two ways. The first is to offer better support for multipoint-to-multipoint communications through the use of multicast (see Deering and Cheriton, 1990, for the seminal paper on the topic). In the current Internet architecture, when a source sends a packet to multiple receivers, the source must replicate the packet and send one to each receiver individually. This results in the several copies of the same packet traversing those links common to the delivery paths (i.e., those links that lie on more than one delivery path, where the delivery path is the route taken by the packet from source to receiver). In multicast, the source merely sends the packet once, and the packet is replicated by the network only when necessary (i.e., the packet is transmitted only once on each link, and then is replicated at the split points where the delivery paths diverge and one copy is sent along each outgoing branch). Sources sending to a multicast group use the multicast group address as the packet's destination address. However, multicast addresses are merely logical names, and do not convey any information about the location of the receivers (unlike a unicast address). Computers on a network wishing to receive packets sent to a particular multicast address send a "join" message to the nearest router. The routing algorithm then distributes this information to create the appropriate distribution trees (i.e., trees from every source to every receiver) so that packets sent to the group reach each receiver. There are a variety of routing algorithms that can accomplish this task; these are not reviewed here. Senders are not aware of who is receiving the packets, because the multicast paradigm is receiver-driven. Efforts to standardize and deploy multicast are well advanced; the vitality of the current MBone (Casner and Deering, 1992) attests to the benefits of this technology.

The second extension to the Internet architecture is much more preliminary, and rather controversial. Efforts are underway to extend the Internet's current service offerings to include a wider variety of QoS. The current single class of best-effort service may not be sufficient to adequately support the requirements of some future video and voice applications (although this is a highly debatable point; see Shenker, 1995b). Moreover, offering all applications the same service is not an efficient use of bandwidth; providing a wider variety of QoS allows the network's scarce resources to be devoted to those applications that are most performance-sensitive. There are many ways in which these services could be extended, some as simple as merely providing several service priority levels and/or drop priority levels (see also the discussion in Clark, 1995a, 1995b, for other approaches to such extensions to best-effort service). Offering multiple qualities of service requires some form of incentives, such as pricing, to encourage the appropriate use of the service classes (see Clark, Shenker, and Zhang, 1992; Estrin and Zhang, 1990; Cocchi et al. 1991; Shenker, 1995a; for discussions of these issues).

More radical extensions to the service offerings are also contemplated. A working group of the Internet Engineering Task Force is preparing a proposal to offer several real-time services; a bounded-delay service, in which the network commits to deliver all packets within a certain delay, is an example of such a real-time service. These services are fundamentally different than best-effort services, in that the network is making an explicit and quantitative service commitment, and therefore must reserve the appropriate resources. Such services require admission control procedures, whereby receivers request service (i.e., issue a reservation request) and the network then either commits to the requested level of service (if it can meet the requirements) or denies the reservation request (if the current load level is too high to meet the requirements of the incoming request). In the proposed resource reservation protocol RSVP (Zhang et al., 1993), receivers send their request for service to the network; this request follows the reverse delivery tree toward all relevant sources (a single source if the application is unicast, or to all senders to the group if the application is multicast). Braden, Clark, and Shenker (1994) presented a slightly out-of-date overview of this proposed architecture.

APPENDIX B: RAMSEY PRICES IN A SIMPLE MODEL

This appendix explores the behavior of Ramsey prices in a simple network model. It considers a facility-providing network service charging a price p for every unit of usage and an attachment cost q. It is assumed that there are no usage-dependent costs associated with the facility; for convenience, only nonnegative prices are considered. The user population is a continuum labeled by α, with $\alpha \in [0,1]$. The usage of each user is denoted by x_α. Utility functions of the form $U_\alpha = V_\alpha - px_\alpha - q$ are considered where V_α represents the valuation of usage; users can detach (yielding a utility $U_\alpha = 0$) if prices are too high. The total welfare is given by $W = \int d\alpha V_\alpha$, the total usage by $Y = \int d\alpha x_\alpha$, and the total revenue by $R = pY + q\int d\alpha$, where the integrals run over all attached users.

The functions V_α take the form $V_\alpha = \lambda_\alpha x_\alpha - x_\alpha^2$. Each attached user sets their

$x_\alpha = \frac{(\lambda_\alpha - p)}{2}$. Three cases for the λ_α are considered: (a) homogenous (where all users have the same V_α), (b) heterogeneous (where users have different V_α) without network externalities, and (c) heterogeneous with network externalities (where one user's valuation depends on the number of other attached users).

In the homogeneous case, $\lambda_\alpha = 1$ is set for all α. The total welfare is $W = \frac{(1-p^2)}{4}$

as long as $0 \leq p \leq 1$ and $0 \leq q \leq \frac{(1-p^2)}{4}$ (otherwise all users detach and $W = 0$).

Thus, welfare is maximized when $p = 0$, $q \leq \frac{1}{4}$ and each $x_\alpha = \frac{1}{2}$. Setting $p = 0$

and $q = \frac{1}{4}$ raises maximal revenue in this case. In this homogeneous case, attachment prices are indeed the optimal way to raise additional revenue.

Now consider a heterogeneous case where $\lambda_\alpha = \alpha$. For a given p and q, all users with $\alpha > A(p,q) = p + 2\sqrt{q}$ remain attached. The total welfare is given by

$$W = \frac{1}{12}(1 - A^3) - p\frac{2}{4}(1 - A),$$

and the total revenue is given by

$$R = q(1 - A) + \frac{p}{4}(1 - A^2) - p\frac{2}{2}(1 - A).$$

The total welfare is maximized when $p = 0$ and $q = 0$. The curve of Ramsey prices—the points that maximize the revenue R—is given by $q=p^2$ for $0 \leq p \leq .2$. The quadratic nature of the Ramsey curve means that increases in usages price dominate (over increases in attachment prices) close to the origin.

Network externalities are introduced by allowing the constants λ_α to depend on the number of other attached users. Let $\lambda_\alpha = \alpha(1-A)$ where, as earlier, A is the critical value of α, such that all users with $\alpha > A$ are attached and no users with $\alpha < A$ are attached. Then $A = \frac{1}{2}(1-(1-8\sqrt{q} - 4p)^{0.5})$. In addition, $W = (1-A)\frac{(1-A^3)(1-A)}{12} - \frac{p^2}{4}$ and $R = (1-A)\frac{(1-A^2)}{4} - \frac{p^2}{2} + q$. The point $p=0$ and $q=0$ maximizes W. The point $p=0.16$ and $q=0$ maximizes R. The Ramsey prices fall along the line segment between these two points: $q=0$ and $0 \leq p \leq 0.16$. For this model, increasing revenue is best done through increasing usage charges only.

The simple model considered here is extremely unrealistic, and neglects important aspects of the problem, such as congestion. However, it does illustrate the basic point that when one has a heterogeneous population containing users who derive marginal benefit from attachment, raising the attachment prices alone is not necessarily a Ramsey price.

APPENDIX C: CAPACITY FILTERS

The ability to express the capacity in terms of an arbitrary filter provides substantial flexibility for accommodating user needs. This appendix gives a concrete example of a sophisticated usage filter. A token bucket filter is parameterized by a rate r and a bucket size b. Usage complies with this capacity as long as the cumulative number of units sent in any time interval of length t (for any such t) is bounded above by $rt + b$. This allows bursts of size b, but bounds the long-term average to be no greater than r. A filter might be the composition of three different token buckets: one with $b = 0$, one with $b = \infty$ (in actual practice, this value of b will be chosen to be large but finite because an infinite sized b imposes no constraints), and one with an intermediate value of b_i. The values of r associated with the two extremal bucket sizes control different aspects of the traffic: r_0 describes the allowable peak rate and r_∞ describes the allowable long-term average rate (the actual size of the large but finite b value used in practice determines, along with the associated rate—the time interval over which this long-term average is applied). The intermediate parameters r_i, b_i describe some intermediate allowable burst rate and size. These different filters should be thought of as constraining the flow on different time scales: the bigger the b, the longer the time scale. In general, one might describe a filter as a nonincreasing function $b(r)$; for every r there is a token bucket with parameters r, $b(r)$ applied to the flow. By adjusting these parameters appropriately, one can provision the capacity for various levels of web browsing or video consumption.

REFERENCES

Braden, R., Clark, D., and Shenker, S. (1994). *Integrated services in the Internet architecture: An overview.* Technical report, IETF. RFC 1633.

Bradford, R. (1994). *Incentive compatible pricing and routing policies in multi-server queues.* Technical Report MS L-195, Systems Research Group, Lawrence Livermore National Laboratory.

Brownlee, N. (1994). New Zealand experiences with network traffic charging. *ConneXions*, 8(12).

Casner, S. and Deering, S. (1992). First IETF Internet audiocast. *Computer Communications Review*, 22, 92–97.

Clark, D., Shenker, S., and Zhang, L. (1992). Supporting real-time applications in an integrated services packet network: Architecture and mechanism. In *Proceedings of Sigcomm 92*.

Clark, D. (1995a). *Adding service discrimination to the Internet.* Technical report, MIT.

Clark, D. (1995b). *A model for cost allocation and pricing in the Internet.* Techncial report, MIT.

Cocchi, R., Estrin, D., Shenker, S., and Zhang, L. (1991). Study of priority pricing in multiple service class networks. In *Proceedings of Sigcomm '91*.

Cocchi, R., Estrin, D., Shenker, S., and Zhang, L. (1993). Pricing in computer networks: Motivation, formulation, and example. *Transactions on Networks*, 1(6), 614–627.

Deering, S. and Hinden, R. (1995). *Internet protocol, version 6 (ipv6) specification.* Technical report, IETF. Internet draft.

Deering, S.E., and Cheriton, D.R. (1990). Multicast routing in datagram internetworks and extended LANS. *ACM Transactions on Computer Systems*, 8(2), 85–110.

Edell, R.J., Mckeown, N., and Varaiya, P.P. (1995). Billing users and pricing for TCP. *IEEE Journal on Selected Areas in Communications*, 13, 1162–1175.

Estrin, D. and Zhang, L. (1990). Design considerations for usage accounting and feedback in intranetworks. Computer Communications Review, 20(5), 56-66.

Francis, P. (1994). Comparison of geographical and provider-rooted Internet addressing. In *Processing of INET '94.*

Gawlick, R. (1995). *Admission control and routing: Theory and practice.* Technical report, MIT.

Gon, J. and Srinagesh, P. (1995). *The economics of layered networks.* Technical report, Bell Communications Research, Inc.

Gupta, A., Stahl, D.O., and Whinston, A.B. (1994). *Managing the Internet as an economic system.* Technical report, University of Texas at Austin.

Gupta, A., Stahl, D.O., and Whinston, A.B., (1995). *The Internet: A future tragedy of the commons.* Techncial report, University of Texas at Austin.

Herzog, S., Shenker, S., and Estrin, D. (1995). Sharing the "cost" of multicast trees: An axiomatic analysis. In *Proceedings of Sigcomm '95.*

Jiang, H. and Jordan, S. (1995). The role of price in the connection establishment process. *European Transactions on Telecommunications and Related Technologies.*

Jordan, S. and Jiang, H. (1995). Connection establishment in high speed networks. *IEEE Journal on Selected Areas in Communications, 13,* 1150–1161.

Li, Y.-J. and Hui, J. (1994). Service shadow price and link shadow price. In *Proceedings of the 2nd International Conference on Telecommunications Systems Modeling and Analysis.*

MacKie-Mason, J.K., Murphy, J., and Murphy, L. (1995a). *The role of responsive pricing in the Internet.* Technical report, University of Michigan, Dublin City University and university of Auburn.

MacKie-Mason, J.K., Murphy, J., and Murphy, L. (1995b). *ATM efficiency under various pricing schemes.* Technical report, University of Michigan, Dublin City University and University of Auburn.

MacKie-Mason, J.K., Shenker, S., and Varian, H. (1995). *Service architecture and content provision.* Technical report, University of Michigan and Xerox PARC.

MacKie-Mason, J.K. and Varian, H. (1995a). Pricing the Internet. In B. Kahin and J. Keller (Eds.), *Public access to the Internet.* Englewood Cliffs, NJ: Prentice-Hall.

MacKie-Mason, J.K. and Varian, H.R. (1995b). Pricing congestible network resources. *IEEE Journal on Selected Areas in Communications, 13,* 1141–1149.

Markoff, J. (1993, November 3). Traffic jams already on the information highway. *New York Times,* p. A1.

McLean, R.P. and Sharkey, W.W. (1993). *An approach to the pricing of broadband telecommunications services.* Technical report, Bellcore.

Mohring, H., and Harwitz, M. (1962). Highway benefits: An analytical approach. Evanston, IL: Northwestern University Press.

Murphy, J. and Murphy, L. (1994). Bandwidth allocation by pricing in ATM networks. In Proc. of *IFIP Broadband Communications '94*, Paris, France.

Murphy, L. and Murphy, J. (1995). Pricing for ATM network efficiency. In *Proc. 3rd International Conference on Telecommunication Systems Modelling and Analysis*, Nashville, TN.

Parris, C. and Ferrari, D. (1992). *A resource based pricing policy for real-time channels in a packet-switching network*. Technical report, International Computer Science Institute, Berkeley, CA.

Parris, C., Keshav, S., and Ferrari, D. (1992). *A framework for the study of pricing in integrated networks*. Technical report TR-92-016, International Computer Science Institute, Berkeley, CA.

Shenker, S. (1994). Making greed work in networks: A game-theoretic analysis of switch service disciplines. In *Proceedings of Sigcomm '94*.

Shenker, S. (1995a). Service models and pricing policies for an integrated services Internet. In B. Kahin and J. Keller (Eds.), *Public access to the Internet*. Englewood Cliffs, NJ: Prentice-Hall.

Shenker, S. (1995b). Some fundamental design decisions for the future Internet. *IEEE Journal on Selected Areas in Communications, 13*, 1176–1188.

Srinagesh, P. (1995). *Internet cost structure and interconnection agreements*. Technical report, Bellcore.

Stahl, D.O. and Whinston, A.B. (1992). *An economic approach to client-server computing with priority classes*. Technical report, University of Texas at Austin.

Strotz, R.H. (1965). Urban transportation parables. In J. Margolis (Ed.), *The public economy of urban communities* (pp. 127–169). Washington, DC: Resources for the Future.

Wang, Q., Sirbu, M.A., and Peha, J.M. (1995). *An optimal pricing model for cell-switching integrated services networks*. Technical report, Carnegie Mellon University.

Zhang, L., Deering, S., Estrin, D., Shenker, S., and Zappala, D. (1993). RSVP: A Resource ReSerVation Protocol. *IEEE Network Magazine*, 7, 8–18.

III

INTEROPERABILITY

8

The Transition to Digital Television Distribution Systems: A Technological View of Expected Interoperability*

David P. Reed
*Cable Television Laboratories,
Louisville, CO*

With the deployment of two digital direct broadcast systems (DBS) last year, the transition to digital television systems to distribute entertainment video services began in the United States. With still more DBS launches on the horizon, cable-TV networks will likely be the next major distribution system to begin widespread deployment of digital video compression technology next year. Over-the-air broadcasters and wireless cable providers are likewise formulating plans to deploy digital systems over their networks in the near future, as are the telephone companies. Indeed, the anticipation of a highly competitive market for digital distribution systems is accelerating the roll out of these systems by all providers.

This chapter describes the current technological status of digital television distribution systems in the United States at the end of the summer of 1995. Given that this technology is now in the early development and deployment stage, such a "snapshot" of the current situation suffers from some lack of documented technical information on different systems. Despite this drawback, however, this chapter is able to identify and discuss several similarities and differences among the existing and proposed digital television compression systems of DBS providers, cable-TV operators, and over-the-air broadcasters. Such a comparison of system characteristics provides a useful inference of the degree of interoperabil-

---

* The views expressed in this chapter are solely those of the author and do not necessarily reflect those of Cable Television Laboratories (CableLabs) or its member companies.

127

ity that might be reasonably expected across the different distribution systems over time.[1]

To this end, the next section begins by providing a technological overview of digital television distribution systems. The following section compares the technical characteristics of various digital television distribution systems. The final section concludes the chapter with a brief discussion of the degree of interoperability expected across digital television systems.

TECHNOLOGICAL OVERVIEW
OF DIGITAL TELEVISION DISTRIBUTION SYSTEMS

The technological components of digital television distribution systems often are described as a hierarchy of layers, where the purpose of each layer is to offer certain services or functionalities to higher layers in the hierarchy. Table 8.1 illustrates one simple model for digital television distribution systems consisting of four layers, along with a description of the functions performed in each layer. The bottom three layers of the model represent the system components that are related to the distribution network. The presentation layer defines the generic technical requirements of production and consumer electronics video equipment.

TABLE 8.1
Hierarchy of Layers of a Digital Television Distribution System

Layer	Description
Presentation	Picture and sound formats for raw video and audio samples
Compression	Transforms samples to/from coded bit stream
Transport	Performs packetization, multiplexing, and synchronization of bit streams
Transmission	(De)Modulates bit stream for transmission; performs error correction

The remainder of this section describes some of the salient technological aspects and concerns for each of the layers described in Table 8.1. This discussion is not intended to provide a comprehensive technical overview of each layer of any given digital television distribution system.[2]

---

1. Throughout this chapter *interoperable* systems are those that can operate together in their native, or untranslated, format. *Compatible* systems, in contrast, are those that operate together with the assistance of equipment that translates between different formats.

2. For a more technical discussion of the topics discussed later, see the ISO/IEC 13818-1 MPEG-2 Systems Standard; ISO/IEC 13818-2 Video Standard; and ISO/IEC 13818-3 Audio Standard; Singapore, Nov. 1994. See also, "The Grand Alliance: The U.S. HDTV Standard," *IEEE Spectrum Magazine*, April 1995, pp. 36-45; and "Multiple Service Transmission via the Grand Alliance System," Appendix G, Advisory Committee Final Report and Recommendation, FCC Advisory Committee on Advanced Television Service, November 28, 1995.

Presentation Layer

The presentation layer includes the scanning formats for the television picture and the accompanying audio. Table 8.2 lists the picture formats under consideration for digital television systems.

With regard to picture quality, important parameters of the format include the number of vertical active lines in the picture, the number of active samples per line in the picture, and the frame or picture rate (how often the picture is refreshed). Clearly, the larger the number of active vertical lines and active samples per line in a television picture, and the higher the picture rate, the better the picture quality or higher the picture resolution. Today's analog televisions have a resolution defined by 483 active vertical lines and 440 active samples per line.[3] Looking at Table 8.2, one can see that the resolution of high-definition (HDTV) pictures of digital television are much higher than today's television, whereas standard definition (SDTV) digital television pictures may be the same or much better (independent of the picture resolution format, SDTV will often look better than today's analog pictures because most analog pictures will include noise impairments, whereas digital pictures will not).

TABLE 8.2
Existing Analog and Expected Digital Television Scanning Formats

Format	Active samples/line	Active vertical lines	Aspect ratio		Picture rate (Hz)
Today's NTSC	440	483	4 × 3		60I, 30I, 24I
Studio NTSC	720	483	4 × 3		60I, 30I, 24I
SDTV	352,528,640,704	480	4 × 3	16 × 9	60I, 60P, 30P, 24P
HDTV	1280	720		16 × 9	60P, 30P, 24P
	1920	1080		16 × 9	60I, 60P, 30P, 24P

Two other parameters of interest include the aspect ratio and the picture rate. A 16 × 9 aspect ratio will look similar to a movie screen picture, rather than today's 4 × 3 aspect ratio of current television. In addition, an aspect ratio that is equal to the ratio of active samples per line to the active vertical lines also permits square pixels, common to most computer display formats used today. This commonality in formats presumably would facilitate low-cost interoperability between television and multimedia applications.

Looking at the picture rate, programs that originate on videotape use a 60 Hz refresh rate, whereas programs that originate on film use the 24 Hz and 30 Hz rates. If interlaced (I) picture scanning is used, a picture is refreshed by every other

---

3. Actually today's analog National Television System Committee (NTSC) scanning format includes 525 lines, but the 42 other lines in the vertical blanking interval are not used to send picture elements.

line every other frame. In contrast, progressive (P) scanning refreshes a picture by updating each line in succession, which doubles the information transfer rate required over any given time to update a picture relative to interlace scanning. Consequently, the lower picture rates of 30P and 24P can be used to implement progressive scanning schemata at equivalent bit rates to the 60I interlaced scanning formats (and, conversely, due to the high bit rates required with the 60P format, it is unlikely that it will be implemented anytime in the near future because of the high costs required in the decoder). Similar to square pixels, the computer industry favors progressive scanning because this is the scanning technique used in computer monitors, and thus would facilitate interoperability between television and multimedia applications.

Compression Layer

The compression layer transforms the raw video and audio samples into a coded bit stream in preparation for transmission, as well as transforming the coded bit stream back into the raw video and audio samples at the television receiver. Compression of video and audio signals are possible because they contain redundant information. Fundamentally, compression techniques represent a compromise; as the bit rate of the transmitted video or audio signal declines, the compression requires more complex coding algorithms and the picture or audio sound declines in quality. Yet using digital signal processing techniques with inexpensive integrated circuits, state-of-the-art compression techniques can achieve better than 20:1 levels of compression without significant degradation of picture or sound quality.

With regard to compression of digital video signals, an overwhelming international consensus has emerged over the past year to use the Motion Picture Experts Group (MPEG-2) standard. The MPEG-2 standard is an International Standards Organization (ISO) standard for video and audio compression; it also specifies elements of how the compressed data streams are combined in a single transport stream. The MPEG-2 standard describes a system of syntax that, when specified, defines the limits for decoder operation in the receiver and transmission bit rate, and hence picture quality.

Specifically, with MPEG-2, a network provider can define the limits of picture quality for a decoder through specification of the *profile and level* of the compressed signal.[4] Relevant MPEG-2 profiles, in ascending order of potential picture quality, include simple and main profiles. Relevant MPEG-2 levels, again in ascending order of potential picture quality, include low, main, and high levels. Overall, the MPEG-2 profile and level will define the limits of television picture

---

4. The profile identification specifies the subset of MPEG-2 syntax—regarding coding latency, resolution, and scalability—applicable to the bit stream. The level defines a specific set of constraints imposed on parameters in the bit stream, and determines the maximum data rate it can support.

quality of the system. For example, at equivalent bit rates, main profile at main level (MP@ML) decoder provides better picture quality than a simple profile at main level (SP@ML) decoder. However, it is possible for a SP@ML decoder operating at a higher bit rate (usually about 20% greater, although this may vary depending upon the picture scene) to provide the same quality as a MP@ML decoder operating at a lower bit rate. Furthermore, MPEG-2 decoders defined at some compliance point (i.e., the profile and level) can decode images created to be decoded by lower order compliance points within the MPEG-2 syntax. In other words, a MP@ML decoder should always be able to decode a signal encoded at the lower order SP@ML compliance point, but not vice versa.

Why is an understanding of all this seemingly arcane syntax important? The principal reason is that the cost of the MPEG-2 decoder varies significantly depending on the highest order compliance point it can decode. For example, a MP@ML decoder requires more dynamic random access memory (DRAM) than a SP@ML decoder.[5] More specifically, the MP@ML decoder requires 2 Mbytes of DRAM, as compared with 1 Mbyte of DRAM for a SP@ML decoder—a difference in memory requirements that will likely result in an additional wholesale cost of up to $30–$40 per decoder.[6]

To avoid the potentially high costs of MP@ML decoders, yet still gain some of the higher quality pictures provided by these decoders, cable and DBS operators are considering other additional options (called *extensions*). One option, known as "dual prime," is part of the MPEG-2 standard; other option, known as "Digicipher II," is a (currently) proprietary coding system developed by General Instrument (GI). Both sets of extensions would provide better picture quality than a simple SP@ML decoder at equivalent bit rates. It should be noted, moreover, that dual prime and Digicipher II decoders are also expected to decode all MPEG-2 bit streams (e.g., SP@ML) that do not require more than 1 Mbyte of DRAM to decode.

A final level of uncertainty regarding selection of compression layer parameters arises due to the rapid advancement of digital encoders. Recently, MPEG-2 encoders have improved considerably in their ability to deliver better picture quality at any given bit rate. This trend is expected to continue for the foreseeable future as well. Thus, it is conceivable that the picture quality of a SP@ML encoder 5 years

---

5. This cost difference arises because MP@ML compression format uses "B-frames" while the SP@ML decoder probably would not. B-frames improve the level of compression by predicting the current frame from past and future frames. The price of this improvement is a more complex decoder and additional delay in the system to decompress the signal.

6. The spot price for a 0.5 Mbyte DRAM chip recently fell to about $15 from a high near $20 this year. Today there are a limited number of suppliers that sell the DRAM chips required for MPEG-2 decoders, although more suppliers are expected to come online next year. This situation, weighed against the rising demand for DRAM chips from the computer industry, makes any forecast of the anticipated cost of DRAM chips highly uncertain.

in the future will more closely resemble the output of a MP@ML decoder today, despite the use of less memory.

With regard to audio coding, most digital systems in North America are expected to use the Dolby AC-3 algorithm for audio compression which provides 5.1 channels of surround sound at a total bit rate of 384 Kbps.[7] This is currently a departure from the MPEG-2 standard, which specifies the use of the MUSICAM audio system, although the Dolby AC-3 system could be included as an acceptable mode of audio in this standard.

Overall, this discussion illustrates the important elements of the compression layer based on the MPEG-2 standard syntax. Most notably, there is a trade-off among decoder cost, compressed video data rate, and picture quality. To achieve high picture quality, network providers can increase either the complexity, and thus the cost, of the decoder or the bit rate. The best solution to this trade-off will likely vary depending on the unique characteristics of the network providers' network and customers. That is, the optimum compressed video bit rate for each network provider will be different depending on the amount of transmission bandwidth available in the network, the nature of the transmission environment (e.g., there is much less noise in terrestrial cable compared with over-the-air broadcasting), and the nature of consumer demand for picture quality (e.g., how much value do consumers attach to a difference in picture quality between MP@ML and SP@ML?). Moreover, this trade-off assessment must also consider the rapid improvement in digital encoders, which could rapidly improve the output of low-cost decoders in the future.

Transport Layer

The main functions of the transport layer include packetization, multiplexing, and synchronization of the digital signal. As with the previous compression layer, the MPEG-2 standard does provide some definition of how these data streams are transported and managed in a packetized time division multiplexed format. As is discussed later, however, the MPEG-2 standard does not include key management mechanisms of the transport layer.

The MPEG-2 standard calls for the use of fixed-length packets, including a packet identification (PID) in the packet header.[8] The PID identifies the location of the different video, audio, and auxiliary bit streams in the overall transport stream, and provides the mechanism for multiplexing and demultiplexing the elementary streams. The MPEG-2 standard also specifies the format of the program-specific information (PSI) tables that provide the decoder with

---

7. The five channels are left, center, right, left surround, and right surround. The 0.1 channel refers to low-frequency enhancements (the subwoofer).

8. The packet length is 188 bytes, including a 4-byte header.

information about the contents of the transport stream. The PSI tables provide the structure for management of the programs within the transport stream, such as pointers that enable a decoder to extract an individual program selected by a user. However, PSI tables only describe the location of elementary data streams within the transport stream in which they are located (e.g., one 6 MHz channel). Thus, a more global tool beyond the PSI tables is necessary to manage the contents of an entire network.

Service information (SI) tables are the management mechanism designed to meet this need. SI tables describe the location and content of all the transport streams, including the program-related information necessary to operate interactive electronic program guides. Because of the wide disparity in physical parameters of different networks, however, the MPEG-2 standards body recognized that specification of the SI tables were beyond the scope and mandate of the group.[9] Consequently, the MPEG-2 standard only includes a skeletal structure or empty shell for the insertion of SI tables, and not a detailed description.

There are several different SI table formats under development. These include tables by GI, the Grand Alliance, the Digital Video Broadcast effort in Europe, and the two DBS systems operating in the United States (Primestar and DirectTV). The tables developed by GI are generally regarded as being the furthest along for terrestrial systems. Although GI is a member of the Grand Alliance, the Grand Alliance has yet to adopt an SI table format for its system.

Another issue pertinent to the transport layer is the question of whether to use constant or variable bit-rate coding. Constant bit-rate (CBR) coding means that bit rate of the compressed video signal is constant independent of the material in the picture. As a result, CBR coding may have variable picture quality, depending on the complexity of the scene. Variable bit-rate (VBR) coding allows the compressed video bit rate to vary within some range of values. In this way, the quality of the picture can be held relatively constant, independent of the content. VBR in conjunction with statistical multiplexing could further improve efficiency of bandwidth usage, particularly with increasing numbers of channels multiplexed together. Again, however, the choice to use CBR versus VBR may not be universal across all distribution systems, due to the availability of bandwidth and the nature of the services provided.

Finally, although the topic is not covered in this chapter, conditional access systems can be implemented at the transport layer. Each individual transport stream can be encrypted, and the PSI tables can carry encryption and conditional access information needed by the decoder to access the appropriate transport stream. Another approach could be to carry encryption and conditional access information in the SI tables to access any particular transport stream.

---

9. Network-specific information likely to be found in SI tables includes the location of a transport stream by frequency or satellite position, as well as the type of modulation and error correction used.

Transmission Layer

The transmission layer provides the transmission and channel coding portion of the digital video distribution system. This includes the modulation format and the error correction techniques used to deliver the transmitted signal to the receiver. As becomes apparent with the next discussion, the type of transmission system selected by network providers is closely linked to the specific characteristics of each distribution system.

Table 8.3 illustrates several of the modulation alternatives available for digital television distribution systems. How does a network operator choose the appropriate modulation technique for its system? The answer revolves around the availability of the two most common scarce elements of any transmission system: bandwidth and power. If a system is power-limited, such as a satellite system, then modulation techniques that are very robust to noise impairments, but less efficient in use of bandwidth, are appropriate. In contrast, if a system is bandwidth-limited, such as a cable system, then modulation techniques that are very efficient in the use of bandwidth, but consume more power, are best. Other important considerations include the cost and availability of the equipment in very large-scale integration (VLSI) semiconductor chip sets. Table 8.3 offers an illustration of these trade-offs and the applications best suited for the different modulation alternatives.

TABLE 8.3
General Digital Modulation Alternatives

Modulation type	Bandwidth efficiency	Relative noise immunity	Relative cost	Common applications
Frequency shift keying (FSK) and Binary phase shift keying (BPSK)	1 bit/Hz	Strong	Low	Existing addressable converters over cable return path
Quadrature phase shift keying (QPSK)	2 bits/Hz	Strong	Low	Satellite com.; cable return path for telephony
Quadrature amplitude modulation (QAM)	5–8 bits/Hz (64–256 QAM)	Moderate	Medium	Video, telephony, and high-speed data on cable
Vestigial sideband modulation (VSB)	3–8 bits/Hz (8–16 VSB)	Moderate	Medium	Grand Alliance System
Direct sequence spread spectrum	< 1 bit/Hz	Very strong	?	Cordless and cellular telephone
Orthogonal frequency division multiplexing (OFDM)	1–6 bits/Hz	Strong-Moderate	High	ADSL over telco networks; over-the-air TV broadcasting

A second important function of the transmission layer includes the use of forward error correction (FEC) and interleaving techniques. FEC improves the performance of the transmission system by correcting transmission errors at the receiver that have occurred during transmission. Interleaving improves transmission performance by randomizing the impact of noise bursts, thereby allowing better error correction in the event of a noise disturbance. The costs of these techniques are the increased complexity in the receiver, increased transmission delay incurred, and increased amount of bandwidth usually necessary. The type of FEC coding and interleaving best suited for a digital video system depends on the noise environment, channel bandwidth, and other system characteristics. A multitude of FEC coding and interleaving approaches are available for use in digital video distribution systems.

Table 8.4 lists the types of modulation actually in use, or in an advanced stage of development, in various digital television distribution systems. All of these systems use FEC coding and interleaving, although none uses identical techniques. The figures in Table 8.4 illustrate the trade-offs involved in the development of a digital television transmission system. In particular, note that the ratio of the payload data rate to the total data rate, which is a measure of transmission efficiency, improves substantially as the distribution system environment becomes more protected from noise interference. This enables a 6 MHz channel on a cable system to deliver a payload data rate of at least 27 Mbps, whereas the payload data rates over 6 MHz in over-the-air broadcast and satellite systems are 30% and 75% lower, respectively.

TABLE 8.4
Existing and Proposed Digital Video Transmission Formats

Digital video distribution system	Modulation type	Channel size	Total data rate	Payload data rate
DBS (Digicipher I)	QPSK	24 MHz	39 Mbps	27 Mbps
Broadcasters' ATV System	8-VSB	6 MHz	22 Mbps	19 Mbps
ITU cable standard	64-QAM	6 MHz	30 Mbps	27 Mbps
ITU cable standard upgrade	256-QAM	6 MHz	43 Mbps	39 Mbps

COMPARING VARIOUS DIGITAL TELEVISION DISTRIBUTION SYSTEMS

Using the framework of viewing digital television distribution systems through a hierarchy of layers, one can compare the technical characteristics of existing and proposed digital television distribution systems. Table 8.5 lists a reasonable forecast for some of the technical options under development or operating today for the Advanced Television (ATV) system of television broadcasters, a digital video compression system for cable operators, and existing and future digital television system for DBS operators (using the Primestar system as an example).

TABLE 8.5

Technical Options for Digital Television Distribution Systems

Layer		Broadcasters' ATV options	Cable system options	DBS (Prime-star) today	DBS future options
Presentation	HDTV	1280 × 720 P 1920 × 1080 I	1280 × 720 P 1920 × 1080 I	Not Available	1280 × 720 P 1920 × 1080 I
	SDTV	(640 or 720) × 480I	480x(352 ... 720) I	512 × 480	480 × ? I
	Audio	Dolby AC-3	Dolby AC-3	Dolby AC-2	Dolby AC-3
Compression	Analog TV with Set-top	MP@ML (SDTV)	MP@ML (SDTV) SP@ML (SDTV) Digicipher II	Digicipher I	MP@ML SP@ML Digicipher II
	Advanced Digital TV	MP@HL (HDTV) MP@ML (SDTV)	MP@HL (HDTV) MP@ML (SDTV) SP@ML (SDTV) Digicipher II	NA	MP@HL MP@ML SP@ML
Transport		MPEG-2; SI Tables TBD	MPEG-2; GI SI Tables	Digicipher I	MPEG-2; GI SI Tables ?
Transmission		8 VSB	64/256 QAM	QPSK	QPSK

The entries in Table 8.5 assume consumers' television sets would receive digital signals from the ATV system through either a broadcaster's set-top box, if the consumer has an existing analog television set, or directly from the broadcast signal, if the consumer owns an advanced digital television set. Similarly, consumers' television sets would receive digital signals from a cable system through a set-top box or, if it were to become available, through a "cable-ready" advanced digital television set that permits cable operators to perform conditional access and encryption functions through a decoder interface using a "set-back" box. For DBS systems, the assumption is that DBS customers will always require a set-top box to receive satellite signals, even if they own an advanced digital television set.[10] The next section examines the implications of these differences on interoperability between the systems.

INTEROPERABILITY FROM A LAYERED VIEWPOINT

From the standpoint of interoperability, the view emerging from the comparison of the various digital video systems sketched out in Table 8.5 appears to be one of

---

10. Although a discussion of the impact that digital television distribution systems will have on television sets is beyond the scope of this chapter, the preceding discussion illustrates that some large changes could be in store. In particular, the large number of different digital distribution systems suggests that television sets could evolve to be picture monitors, with the receiver equipment moving to a separate box, unique to each network provider, that would also perform new services functions, as well as traditional set-top box functions.

increasing interoperability over time, as the digital video technology matures and moves out of the early deployment stage. With the deployment of first-generation digital systems, some degree of interoperability among broadcast, cable, and DBS systems would seem likely at the presentation and, possibly, compression and transport layers, and not a possibility at the transmission layer. Over time, it appears that interoperability will be achieved across the presentation, compression, and transport layers of these systems as the technology matures with market experience. The next section explores the reasons for this situation for each layer of the digital video system.

Presentation Layer

A common set of compressed HDTV, SDTV, and audio formats to be delivered over broadcast, cable, and DBS systems would appear to be emerging across these systems. That said, the only possible differences that might arise in the near term might be the use of lower resolution modes for SDTV by cable and DBS. Although this difference should not lead to significant interoperability problems—programs displayed in lower resolution modes will not have the same level of picture quality relative to the high-resolution format—it provides another demonstration of how the flexibility of digital technology permits operators to customize their television distribution systems to their best advantage in a competitive market.

To demonstrate, the SDTV formats of the ATV system are either 640×480 or 704×480, both of which would provide essentially "studio" quality pictures that are significantly better than today's analog television picture. Cable and DBS providers, in contrast, are considering half rate (352×480) and three-quarter rate (528×480) resolution options as well. Use of these lower resolution options could permit more programs to be carried over the network due to the lower bit rates. For their part, broadcasters have chosen high-resolution SDTV formats, which initially will only provide enhanced picture quality to high-end television sets.[11] Whether cable or DBS providers opt for lower resolution modes will be largely dictated by market forces such as the characteristics of demand for new digital-based video services using MPEG-2 compression and the changes in the embedded base of television sets.

---

11. Regardless of whether a television set can actually display an improved picture depends on the video inputs available on the set. For example, even if it is receiving a program in studio quality format, existing sets with baseband video inputs such as RCA Phone Plugs or RF inputs using F-Connectors, will provide a resolution of only about 440×480—or no different than today's NTSC picture. Only those high-end television sets equipped with component RGB or S-video jacks can display a studio quality picture with a higher resolution format.

Compression Layer

The emerging international consensus behind the MPEG-2 standard is improving the likelihood that similar, if not identical, compression systems will be used over broadcast, cable, and DBS systems.

Like the presentation layer, however, there are some differences that may emerge due to economic and technical factors that create incentives for the operators of each system to select noninteroperable compression formats. Two issues in particular could have significant implications for interoperability: the question of what will be the highest MPEG-2 compliance point supported by each system, and digital encoder implementation.

Regarding the first issue, recall that MPEG-2 decoders designed for a lower order compliance point will not be interoperable with the output of a higher order MPEG-2 encoder. At this point in time, it is too early to determine whether interoperability will not be present between systems for this reason. Currently, it does appear that broadcasters—who operate in a bandwidth constrained system—favor the MP@ML format because it can reduce the compressed bit stream by roughly 20% relative to the SP@ML format. Cable operators, in contrast, have more bandwidth available on their systems. In addition to the MP@ML format, they are considering the SP@ML mode of operation because it provides the same level of picture quality as MP@ML, avoids the additional memory costs of MP@ML decoders, and facilitates the lowest cost deployment of first-generation set-top boxes possible. Again, market conditions will be a major factor in the outcome of each operator's decision.

Regarding the second issue, interoperability among MPEG-2 decoders designed for the same compliance point is not guaranteed. What will be necessary is the emergence of standard MPEG-2 encoders for each compliance point, a development that will likely require some time as the market for these devices develops.[12] A key requirement for this transition to interoperable encoders will be to confine upgrades to the encoder, which will save the significant cost of upgrading the much more numerous set-top boxes or advanced digital television sets.

Transport Layer

Interoperability of the different transport layers is still undetermined, but should be resolved within a short time. Although the MPEG-2 standard does provide some degree of commonality among all the systems, it falls short of providing complete interoperability at this layer because this standard does not fully specify, for example, the SI table format. If it turns out that different SI tables are used by

---

12. Cable Television Laboratories has developed a conformance test bed to assist the cable industry in evaluating the conformance of different vendors' products to the MPEG-2 standard.

the different digital distribution systems, an outcome that is viewed as increasingly unlikely, then each system cannot identify the location of different digital video program streams within the total bit stream. Such a situation would require the use of compatibility equipment that could convert from one SI table format to another wherever interconnections of non-interoperable systems were to occur.

Transmission Layer

At the transmission layer, interoperability among the broadcast, cable, and DBS systems is not expected in the near or long term due to the different modulation, FEC, and interleaving techniques used in each system. Already these differences are well marked. For digital modulation, broadcasters have chosen VSB, cable has chosen QAM, and satellite has chosen QPSK. As discussed earlier, each of these different modulation techniques provides the most efficient transmission system for the respective distribution environment. Clearly, interoperability at this layer will only be achieved at a cost equal to the additional compatibility equipment necessary to receive the signals from differently modulated transmission systems.

CONCLUSION

This chapter describes the similarities and differences among the existing and proposed digital television distribution systems of broadcasters, cable operators, and DBS operators. A key conclusion that emerges from this comparison is that the inherent flexibility of digital technology provides network operators a wide degree of latitude in specifying the technical characteristics of their transmission networks. As a result, given the highly competitive environment expected for digital television distribution systems, network operators are designing their networks to deliver video in the most efficient manner possible. This means that there could be no interoperability among parts of these systems, particularly in the early stages of deployment when network operators are racing to deploy digital distribution systems in a highly competitive environment.

Over time, interoperability between digital distribution systems should improve as equipment suppliers refine their systems and achieve interoperability where feasible through licensing and strategic alliances. Likewise, where interoperability is not feasible, compatibility among systems should emerge as equipment is built to bridge the interoperability gaps.

Organizing Interoperability: Economic Institutions and the Development of Interoperability

Christopher Weare
University of Southern California

This chapter examines the role of alternative forms of governance—including markets, integrated firms, joint ventures, standards-setting bodies, and regulation—in achieving interoperability in the computer and telecommunications industries. Many technologies in these industries have a systems nature, in that numerous components are interconnected to produce a unified good. For example, what an end user identifies as a "personal computer" comprises several distinct components, including a central processing unit (CPU), a monitor, memory devices, and systems and applications software. Similarly, sending a fax coast to coast entails a facsimile machine, a local network, a long-distance network, another local network, another facsimile machine, and perhaps a component to store the fax and forward it at a more convenient time. Interoperability exists when two networks (e.g., a long-distance and local carrier) or all components of a single network (e.g., software and hardware) work together seamlessly to accomplish the larger task.

Achieving interoperability is a complex process involving a delicate mixture of cooperation and competition. Actors must cooperate to successfully coordinate the design and operation of complicated, interdependent components. At the same time, competition spurs firms to innovate. It can also induce firms to develop open systems in an effort to attract consumers with the promise of larger networks and easy access to complementary products.[1] However, these forces can conflict because competition between actors hinders cooperative efforts, whereas close cooperation dulls competitive appetites.

---

1. Sun Microsystems, for example, has successfully used an open systems strategy to achieve a strong position in the workstation market. See Garud and Kumaraswamy (1993).

This chapter examines the balance of cooperative and competitive forces in interoperability from an organizational perspective. A wide variety of institutional structures have arisen in market economies as a response to the special problems posed by interoperability. In some cases, firms unilaterally decide to make their components compatible, resulting in an interoperable market. In other cases, firms integrate vertically or horizontally so that they may directly manage the interoperability of system components. They may also coordinate their actions through joint ventures and industry standards-setting bodies. Finally, government regulators play a role in this process. For example, the Federal Communications Commission (FCC) has the task of ensuring that telecommunications markets achieve an appropriate level of interoperability.

Although the socially optimal degree of interoperability in computing and telecommunications technologies is not well understood, there is a general sense that institutional constraints can lead to significant social losses.[2] Consequently, understanding how varied institutions may achieve and maintain interoperability is important for policymaking. Moreover, decision makers need a framework within which to identify cases in which government intervention to promote interoperability is appropriate, and to understand the policy problems that arise with such intervention.

The framework employed in this chapter treats the process of attaining and maintaining interoperability as a contracting problem. The choice of organizational forms is assumed to reflect actors' efforts to minimize the transaction costs incurred in crafting contracts and during contractual performance. This perspective illuminates the role played by different organizational forms vis-à-vis interoperability, and highlights the policy problems that arise when policymakers strive to extend the scope of interoperability.

The chapter proceeds as follows: The first section defines *interoperability* and describes the nature of the contracting problem; the second section introduces the transaction cost framework. The third section applies the transaction cost framework to explain the role played by various organizational forms when all firms have incentives to achieve interoperability. The fourth section considers the policy implications when dominant firms strategically prevent rivals from producing compatible products. Specifically, it identifies easy cases in which rivals or well-defined regulatory interventions can thwart such strategies, and hard cases in which regulatory efforts to expand the scope of interoperability pose fundamental contracting problems that are not adequately addressed by any feasible organizational form. The fifth section concludes.

---

2. Interoperability is not always desirable. For example, there are benefits to technological variety. Consumers' tastes for network features may differ, and the simultaneous development of competing technologies may facilitate innovation. A full treatment of the costs and benefits of interoperability, however, is beyond the scope of this paper.

INTEROPERABILITY AND CONTRACTING

The Computer Systems Policy Project (1994) defined *interoperability* as the ability of two or more systems to interact, using a prescribed method to achieve a predictable result. Interoperability allows diverse components and entire systems made by different vendors to communicate with one another. Interoperable systems are compatible at prescribed levels of interaction, which is achieved through the specification of interfaces between components and systems.

Interoperability has several important dimensions. *Horizontal interoperability* involves compatibility between similar components of the same system, such as two operating systems running on different computers. This type of interoperability affects the scope of networks, as well as the degree and type of competition between rival systems. *Vertical interoperability* involves the compatibility of one component (e.g., a CPU) with complementary components (e.g., software). It determines the variety of complements available for a system and the degree of competition in complementary markets. Finally, *temporal interoperability* involves compatibility between successive generations of a particular technology. For example, temporal interoperability enables software designed to run on the Intel 386 microprocessor to also run on the new Intel Pentium microprocessor.

In all of these aspects, interoperability is a continuum. At one end, full interoperability between systems implies that all functionalities of both systems are available at equal levels of performance; on the other end, total incompatibility implies that one system is completely unable to access the other. Between these extremes is a wide array of possible configurations, in which one system either cannot access certain functionalities or suffers from a significant performance handicap. For example, the PowerPC microprocessor can run software written for Intel x86 microprocessors, such as Microsoft Windows, but it does so more slowly (Chang, 1994).

The process of achieving interoperability between components or systems entails the specification of technical and financial parameters. As such, achieving interoperability can be conceptualized as a contracting problem between actors. Such contracts have to address several central issues, either explicitly or implicitly, including the technical characteristics of interfaces, compensation for research and development and other investments, rules to coordinate changes in technology, and rules to resolve disputes.[3]

These fundamental dimensions of contracting can be addressed in numerous organizational settings, including competitive markets, standards-setting bodies, intrafirm negotiations, complex interfirm agreements (e.g., joint ventures), and regulatory bodies. For example, in a market setting, the technical characteristics of interfaces depend on the independent design decisions of firms, and compensa-

---

3. Existing research on technical compatibility tends to focus on the first part of this contracting problem (agreement on standards) but places much less emphasis on the other dimensions of contracting.

tion is determined purely by market clearing prices. Interoperability is achieved when producers of complementary products respond to market prices and consumer demand by designing their products to be compatible with others available on the market.

On the other end of the spectrum, a single firm that produces a number of system components may closely coordinate all dimensions of contracting. It can establish internal standards to ensure the interoperability of its components, coordinate and compensate investments in research and development (R&D) and new components, and resolve disputes that may arise. In the middle of this spectrum are intermediate organizational forms that combine the features of markets and hierarchies. Standards-setting bodies, for example, address the specification of interfaces and the establishment of rules for amending technologies in a coordinated fashion, while leaving other contractual issues, such as the compensation for investments, to the market.

Government also plays an intermediate role in interoperability through regulation and antitrust enforcement. At times, regulators intervene directly in setting standards. For example, the FCC has mandated specific standards for the broadcast industry, including the National Television Standard Code (NTSC) broadcast standard for color television. More important, regulators create the general legal framework within which firms address contracting issues. For example, in the 1986 Third Computer Inquiry, the FCC required dominant local telephone companies to open their networks to third parties; in the process, it prescribed rules for establishing interfaces, setting prices, and resolving disputes.

Each of these organizational forms approaches the basic dimensions of contracting in a qualitatively different manner. Thus, a better understanding of their costs and capabilities in accomplishing these tasks is necessary to understand their role and analyze possible improvements to current practice. Transaction cost economic theory offers a powerful analytic framework with which to examine these issues.

TRANSACTION COSTS ANALYTICAL FRAMEWORK

The central premise of transaction cost economics, developed mainly by Williamson (1975, 1985, 1991), is that firms in competitive markets organize economic activity to minimize transaction costs. These include the costs of gathering information, negotiating, monitoring contractual performance, adapting contractual terms to changing circumstances, and resolving disputes. The analysis compares how different organizational forms govern the entire contracting process, including *ex ante* incentives crafted in the contract, as well as *ex post* performance of contractual terms. Special emphasis is placed on *ex post* control of opportunistic efforts to extract concessions from trading partners when contractual circumstances change.

Transactional Characteristics

A significant insight of transaction cost economics is that the issues that arise during contractual performance depend on the characteristics of the underlying transactions. Two characteristics are central: asset specificity and uncertainty.[4]

ASSET SPECIFICITY. Asset specificity arises when idiosyncratic investments support a transaction. Such specialized investments are worth significantly less in their next best use. Consequently, firms can only secure the full value of the investment by maintaining existing trading relations.

Asset specificity is a key determinant of organizational choice because of its effect on *ex post* contractual performance. When bilateral dependence arises from specific assets, contracting parties become vulnerable to opportunistic behavior because they cannot costlessly realign their trading relationships. In response, actors have incentives to craft complex contractual safeguards to mitigate these dangers. In contrast, when the absence of specific investments enables parties to switch trading partners at low cost, the threat of exit adequately mitigates the dangers of opportunism.

An important caveat is that fixed investments only require contractual safeguards when they are specific to a particular transaction, rather than a particular industry. Investments in a network component expose a firm to the dangers of opportunism if only one firm produces complements. If numerous suppliers of all network components exist, however, problems of opportunism are mitigated. Thus, technical standards that facilitate competition in component markets reduce transaction-level asset specificity, although industry-level specificity remains.

UNCERTAINTY. Uncertainty arises in highly unique situations where neither all possible outcomes nor their probabilities are known.[5] Rapid technological innovation typically involves such uncertainty because the innovative process is inherently unpredictable, and the consequences of major innovations are rarely appreciated beforehand.

Uncertainty increases the hazards that parties face in the contracting process. Contracts remain incomplete because parties cannot foresee all possible futures or agree on appropriate actions in such contingencies. Consequently, during *ex post* performance, contingencies will arise that are not clearly covered in the contract and that will require mutual agreement to readjust terms. However, such renegotiations can be contentious as parties strategically maneuver to benefit from the new circumstances.

---

4. Williamson (1991) includes other dimensions of transactions in the analysis, such as the frequency of transactions and the underlying contract law regime, that are not considered in this analysis.

5. In contrast, situations in which a future outcome is unknown but there is a known probability distribution involve risk but not uncertainty. See Knight (1965).

Combined, uncertainty and asset specificity create the most pronounced transactional difficulties. In the absence of asset-specific investments, uncertainty does not require more complex forms of governance because actors can easily rearrange trading relations in response to disturbances. Conversely, in the absence of uncertainty, actors can mitigate the dangers of opportunism by crafting complete contracts. When both are present, however, more complex organizational arrangements are needed to resolve ongoing uncertainties while mitigating opportunistic behavior.

Governance

Firms select discrete organizational forms to govern transactions by balancing the strength of economic incentives against the ability to mitigate opportunistic behavior and coordinate adaptation. Spot market transactions involve high-powered incentives, in that actors are the full residual claimant for their actions. The down side is that market relationships are brittle because self-interest can lead to opportunism. Internalizing a transaction within a firm creates low-powered incentives because employees' compensation does not fully reflect their specific outputs or level of effort, but a firm has managerial mechanisms for avoiding opportunism and effecting coordination. Hybrid organizational forms, such as long-run contracts and joint ventures, provide intermediate degrees of incentive intensity and adaptability.

The form of governance that minimizes transaction costs depends on matching transactional characteristics with organizational capabilities and costs. In brief, transaction cost economic theory predicts that market governance is most efficient when transactional complexity is low (e.g., in the absence of uncertainty, coordination requirements, and specific investments). In these cases, markets have the substantial advantages of high-powered incentives and low organizational costs. As transactional complexity increases, however, hierarchical governance (e.g., the integration of transactions within a single firm) becomes more efficient because it more effectively coordinates actions, resolves disputes, and adapts to change. The advantages of integrated governance, however, must be balanced against the added costs of increasing the scope of a firm. Because transactions within a firm are not disciplined by market incentives, firms rely on monitoring and auditing to control performance. Management's ability to undertake such activities, however, decreases as the size and scope of the firm increases. Thus, there are limits to both firm size and the scope of technological regimes that a single firm may govern efficiently (Teece, 1988).

THE GOVERNANCE OF INTEROPERABILITY

Economists have recently made significant contributions to the understanding of how compatibility emerges in noncooperative market settings. A large body of literature examines firms' incentives to achieve interoperability, and how independent decisions by firms and consumers affect the timing and choice of compatibility. Katz and Shapiro (1985) and Economides (1991) showed that firms have private incentives to achieve interoperability if positive network externalities[6] exist or if consumers have high demand for services that span networks. Firms are most likely to agree to achieve interoperability between their products when they all privately benefit from doing so. Even when the costs and benefits of interoperability are distributed unevenly, firms may agree if the overall benefits of interoperability are large and the firms that benefit most are willing and able to compensate firms that are harmed. However, there are circumstances in these models when some firms resist interoperability even if it is socially beneficial. Specifically, firms are less likely to cooperate in developing interoperability if they have large networks and good reputations that provide them with an advantage in attracting consumers, and also when consumers do not strongly demand cross-network services.

Even when proper incentives exist, achieving interoperability depends on the ability of firms and consumers to coordinate technologies. One mechanism to resolve this coordination problem is a bandwagon in which one firm autonomously adopts a technology and others follow to remain compatible. Farrell and Saloner (1985) showed that when there is no uncertainty concerning the preferences of other actors and all actors benefit from compatibility, bandwagons always achieve interoperability. Under conditions of uncertainty, or when payments between firms are not feasible, however, bandwagons may fail to achieve a socially optimal level of compatibility. Another possible coordination mechanism is a standards-setting body, in which member firms directly exchange information. Farrell and Saloner (1988) showed that coordination through standards bodies outperforms bandwagons, but that a combination of both mechanisms outperforms either individually.

Transaction cost analysis complements these more traditional analyses of compatibility and standards. By emphasizing *ex post* contractual performance, it clarifies how the coordination issues that arise in achieving interoperability are conditioned by market and technological circumstances. In addition, the contracting framework emphasizes the organizational consequences of opportunism, contractual disputes, and a rapidly changing environment, all of which are major, recurring issues in the computer and telecommunications markets. In summary, an analysis of transactional characteristics and organizational capabilities can illuminate the circumstances

---

6. Network externalities exist when the benefit derived by a consumer from consuming a network good depends on the number of other individuals connected to the network. A telephone network, which is more valuable the more people that may be reached on it, is a prime example.

under which market-based interactions are likely to lead to interoperability, when nonmarket institutions are necessary, and the issues policymakers are likely to face.[7]

To simplify this analysis, the two dimensions of transactions—asset specificity and uncertainty—are treated as dichotomous variables, that have either a high or a low value. Interoperability involves high asset specificity when investments in compatible components create bilateral relations. Such investments are most likely for new, cutting-edge systems, for which standards have not yet evolved, generic technologies are not suitable, and markets remain thin. In contrast, asset specificity is low when technologies are well defined, interfaces are standardized, and many firms produce complementary components. Uncertainty is high when the preferences of firms controlling complementary components, the relative merit of alternative technologies, or the strength of consumer demand are unknown.

As seen in Figure 9.1, the two dimensions yield four types of transactions for achieving interoperability. Each transactional type poses different contracting issues. Thus, the differential capabilities of governance institutions to address these issues are considered, including the specification of technical parameters, incentive design, and the development of rules for adaptation and dispute resolution. This allows the identification of the form of governance that enables actors to achieve and maintain interoperability while minimizing transaction costs. To emphasize the organizational dimension of interoperability, coordination issues are focused on by assuming that firms have joint incentives to achieve interoperability. In this case, four general forms of governance arise: markets, coordination mechanisms, contracting, and complex governance. The remainder of this section examines these matches; the following section extends this analysis to include the regulatory issues that arise when firms wish to strategically impede interoperability.

Markets

As shown in the first quadrant of Figure 9.1, when asset specificity and uncertainty are low, actors are able to achieve interoperability most efficiently through market-based interactions. In the absence of technological uncertainty, firms are less likely to disagree about the relative merit of alternative technologies, facilitating coordination around a particular standard. Similarly, when market demand is well known, firms will have similar expectations concerning the benefits of interoperability, facilitating coordination and the design and implementation of side payments. In addition, market governance is preferred to more complex forms because, in the

---

7. The contracting implications of different types of transactions are not the sole determinant of industry structure in the computer and telecommunications markets; economies of scale, history, and other factors are also important. In addition, because these industries are particularly dynamic, the organizational structures observed are likely to differ from what would be expected in an equilibrium. Nevertheless, the character of underlying transactions is an important and often overlooked factor in the structure of these industries.

Figure 9.1
Transactional characteristics determine
governance for interoperability

Uncertainty

	Low	High
Low	**1** **Markets** • Autonomous decision making • Bandwagons	**2** **Coordination Mechanisms** • Traditional standards setting • Regulatory standards setting
High	**3** **Comprehensive Contracting** • Complete contracts • Anticipatory standards setting	**4** **Complex Governance** • Integrated governance • Joint ventures • Regulation

Asset Specificity

absence of asset specificity, firms do not develop bilateral relationships and have little need for more complex dispute resolution or adaptive mechanisms.[8]

The introduction of audio CD players demonstrates how market-based decisions can lead to interoperability. Firms could have marketed CD players as stand-alone music systems, or they could have made them incompatible with existing stereo components. Nevertheless, the benefits of making CD players compatible with a large installed base

---

8. Even absent asset specificity, firms are still in danger of having their investments stranded if they adopt one technology and the market standardizes on an incompatible technology after investments have been made. This problem represents a contracting failure because early adopters of a technology are generally unable to contract with later adopters. This problem differs from the contractual hazards that arise from bilateral dependence in that it is purely a coordination issue that does not require the control of *ex post* opportunism.

of stereo equipment were clear, and the specification of interfaces was well known. Thus, firms had both the incentives and capabilities to make CD players interoperate.

An example of a bandwagon coordinating independent firm choices is the adoption of a scrambling device in the cable industry. In the early 1980s, programmers wished to prevent satellite dish owners from obtaining unauthorized viewing of program transmissions. In December 1985, Turner Broadcasting and Showtime announced their choice of a particular technology; within two months, the rest of the industry followed, creating a de facto standard. In this case, several factors facilitated quick adoption. Firms had incentives to employ a compatible technology because it facilitated the marketing of their programs to satellite dish owners. The Turner/Showtime decision removed market uncertainties, and the choice of scrambling technology did not create bilateral dependencies, thus enabling firms to join the bandwagon quickly without fear of future opportunism.[9]

Market-based transactions are also appropriate for technologies governed by reference standards, such as the International Standards Organization's (ISO) Open Systems Interconnection model for computers. Reference standards establish a general framework for the relationships among components in a system. Thus, they reduce technological uncertainty. If several competitors produce products at each level of the model, they also reduce the dangers of bilateral dependence.

Coordination Mechanisms

Movement from the first to the second quadrant of Figure 9.1 increases uncertainty, but leaves asset specificity unchanged. In these cases, interoperability can best be achieved through voluntary industry standards-setting organizations or regulatory mandated standards. Before firms and users are willing to invest in system components, they must resolve uncertainties concerning the preferences of actors, technological capabilities, costs, and consumer demand. As Farrell and Saloner (1985) showed, uncertainty can impede actors from independently adopting a new standard, even if they are made better off by such a move. Consequently, market-based interactions must be augmented with coordination and communications mechanisms to resolve such uncertainties.

A useful mechanism to facilitate this is standards-setting bodies, such as the many committees that operate under the auspices of the American National Standards Institute (ANSI) or the ISO. Studies of these organizations have concluded

---

9. In contrast, if joining a bandwagon requires investments in specific assets, it is less likely to effectively coordinate firm choices; firms will hesitate to follow the market leader unless some safeguards exist for those investments. For example, in telecommunications the adoption of integrated services digital network (ISDN) technology has proceeded slowly. The lack of a strong bandwagon effect can be explained in part by the hesitancy of producers of complementary equipment to invest in specialized R&D and equipment whose value is highly contingent on technical and pricing decisions of producers of complementary products.

that one of their critical functions is enabling parties to exchange information. Besen and Johnson (1986) found that standards-setting bodies are more likely to be successful when they uncover important technical differences between competing technologies. Sirbu and Weiss (1990) found that standards-setting bodies' decisions are significantly influenced by the information provided by participants. Firms sponsoring a proprietary technology in a standards-setting body also frequently make commitments concerning royalty payments, thereby resolving financial uncertainties faced by other firms that may adopt the technology.

A limitation of standards-setting bodies is that they are voluntary. Thus, they cannot enforce their recommendations or mitigate opportunistic behavior. This constraint, however, is unimportant for these types of transactions because, in the absence of specialized investments, firms do not require contractual safeguards to be willing to invest in complementary components.

Regulatory standards setting is similar to voluntary industry bodies, in that it is principally a method for gathering information. From this perspective, it appears inferior to voluntary bodies because regulators must rely on third parties that have incentives to distort information, and they are hampered by a lack of technical expertise. However, mandated standards have the advantage of being able to enforce agreements. When agreement on a particular technology requires compensating some firms, market uncertainties can impede the crafting of side payments. Regulators, in contrast, avoid the need to reach such a consensus because standards set by regulators have the force of law. The case of AM stereo is instructive. In 1982, the FCC left the development of a standard to the market. Over the next 10 years, however, no single standard emerged because conflict between equipment suppliers precluded voluntary industry agreement. Consequently, the technology diffused slowly, and, in 1992, Congress ordered the FCC to mandate a standard.

Comprehensive Contracting

The third quadrant in Figure 9.1 combines high asset specificity with low uncertainty. Under these conditions, interoperability can best be achieved through the use of comprehensive, contingent contracts. To be willing to place investments in complementary assets that lead to bilateral relationships, firms require safeguards to protect against opportunism. Because uncertainty is low, however, firms are able to craft sufficient safeguards in contracts, thereby avoiding the need for more complex and expensive forms of governance.

The market for telephone switching equipment is governed by such contracts. Traditionally in Europe, and since the divestiture of AT&T in the United States, the major switch manufacturers have been separate from telephone carriers. Nevertheless, these firms exist in close bilateral relationships. Network interoperability depends critically on switch control software. Subtle differences between competing switches lead to incompatibilities that greatly limit the ability of carriers to switch suppliers, and thereby create contractual hazards. To address such hazards, at least in part, carriers develop highly detailed technical specifications for equipment and test prototypes before purchase. For example, Bellcore performs these duties for the seven Regional Bell Operating Companies (RBOCs).[10]

Another form of governance appropriate for these types of transactions is anticipatory standards setting. As the name implies, the establishment of standards precedes the development and marketing of components. The effort by the Consultative Committee for Telephone and Telegraph (CCITT) to develop standards for ISDN is a prime example. The combination of high asset specificity and low uncertainty makes anticipatory standards attractive and feasible. They are attractive because the establishment of standards reduces contractual hazards by promoting the interoperability of numerous firms' products.[11] They are feasible when uncertainty is sufficiently muted because functionalities can be rigorously described *ex ante*, new technologies do not threaten to make the standard obsolete before it is implemented, and the standards correspond to services demanded by consumers.

Complex Governance

Transactions found in the fourth quadrant of Figure 9.1 are characterized by both high uncertainty and asset specificity, and they require complex forms of governance. In such cases, actors require safeguards to invest in complementary system components. Moreover, frequent, unexpected changes in market and technological circumstances require mechanisms that enable actors to adapt continuously while mitigating opportunism, resolving disputes, and preserving existing trading relationships. Because of the degree of uncertainty involved in such cases, however, it is difficult to craft comprehensive contracts *ex ante*. The natural response to these contractual hazards is to internalize such transactions within a firm. Integrated governance allows these transactions to operate with a greater degree of contractual incompleteness while controlling opportunism. Centralized control also facilitates the communication of tacit knowledge and the coordination required for the adaptation of systems components to changing technology.

Integrated governance is not always feasible, however, because of antitrust constraints or limits on the scope of a firm. For example, the 1982 AT&T consent decree prohibits the RBOCs from providing long-distance services, even though

---

10. Recent efforts by the Bell Operating Companies to be allowed to enter the equipment manufacturing market may indicate that transactions for switches and other major types of equipment may involve sufficient uncertainty that comprehensive, contingent contracts are not an efficient form of governance. Nevertheless, MCI, Sprint, and European carriers still rely on contracting.

11. Such standards, however, do not guarantee interoperability, as was seen in the first effort to develop ISDN in the United States.

there are systemic innovations in the telecommunications network that may be more efficiently managed through centralized governance.

Interoperability issues can also extend beyond the bounds of a single firm when the components of a system are based on disparate technologies. For example, the development of network-based multimedia applications requires expertise in computing, telecommunications, graphics, and entertainment. In these situations, integration poses hazards because a single firm is not well equipped to monitor the performance of novel functions.[12]

Joint ventures are an attractive alternative when integration is not feasible. They mitigate the dangers of opportunistic behavior by aligning incentives through the commingling of assets and profit sharing. Moreover, joint ventures promote the exchange of hard-to-communicate information by bringing employees with disparate expertise into direct contract with one another. Joint ventures are still susceptible to breakdown, however, when significant changes in market circumstances give one partner incentives to abrogate the agreement unilaterally.[13] Williamson (1991) argued that, in the long run, they are likely to be replaced by structures that are more or less integrated, depending on underlying transactional characteristics.

Summary

This organizational analysis of the issues that arise in establishing and maintaining interoperability in the telecommunications and computing industries leads to three important insights. First, the character of transactions matters. Each of the four transactional types examined earlier leads to a particular mix of contracting issues. In turn, these are most efficiently addressed by distinct forms of governance. A good deal of the variation in organizational forms observed in these industries can be understood as efficiency-based responses to different degrees of uncertainty and asset specificity in the underlying transaction.

Second, there is a reciprocal relationship between governance and interoperability. Clearly, the governance structure that best facilitates interoperability closely depends on transactional characteristics. A more subtle point is that causality also operates in the reverse direction. The organizational forms chosen in one period influence future interoperability and standards, which, in turn, shape the optimal

---

12. A vivid example of these problems is Sony's purchase of Columbia Pictures. Sony was motivated to purchase a Hollywood studio to assure that compatible software would be available for its entertainment hardware. Lax monitoring, however, led to poor management and forced Sony to accept a multi-billion dollar write off in 1995.

13. Microsoft and IBM, for example, announced an agreement in 1989 under which they would jointly migrate software development from Microsoft's Windows operating system to IBM's OS/2 system. In the next year, Microsoft broke the agreement. Most likely, Microsoft's change of heart came about as it realized that Windows was going to dominate the market.

organizational form in later periods. For example, if interoperability is achieved through the specification of interface standards, technical uncertainty is reduced and firms are freed from bilateral dependence because large numbers of firms can produce complementary components. Thus, there is a tendency for transactions to gravitate, over time, toward market-based governance once standards are created and markets become more established.

Finally, when firms have incentives to produce compatible components and systems, as has been assumed to this point, interoperability does not pose a significant policy problem. As the analysis of governance has shown, private-sector forms of governance adequately address the various coordination and contracting problems that arise in the development of interoperability.[14] When uncertainty and asset specificity are present, the efficient development of interoperable components does depend on more complex private-sector institutions. Thus, policymakers should generally tolerate vertical integration and other nonconventional business arrangements.

For example, lax antitrust review of joint ventures is warranted, despite the danger that joint ventures can facilitate collusion when they involve a significant portion of all firms in a particular market. Economides and White (1993) pointed out that multibank ATM networks create a vehicle by which the banks can fix prices for a broad range of banking services or discipline maverick service providers. Nevertheless, the benefits of coordination appear substantial. Thus, it appears preferable to police the joint venture for inappropriate activities, rather than forbid it outright.

STRATEGIC USES OF INTEROPERABILITY AND REGULATORY POLICY

In this section, the analysis from the previous pages is extended to examine situations in which firms have incentives to impede compatibility even though social welfare may be increased with greater interoperability. Firms with a large installed base relative to their rivals and dominant market positions will strive to maintain market power by retaining proprietary control of interfaces and impeding rivals from producing compatible components (Greenstein, 1990; Adams & Brock, 1982). When a firm builds a large installed base of its proprietary technology, it

---

14. Private sector institutions do not, however, attain interoperability under all circumstances. Markets, characterized by increasing returns, tend to irreversibly follow one technological trajectory once it is chosen. Moreover, the choice of trajectory is very sensitive to chance events, the preferences of key decision makers, and even personalities. Thus, it is easy to identify markets in which either an inferior technology dominates (e.g. VHS over Beta) or a suboptimal degree of interoperability exists (e.g. the mainframe computer market). Analysts point to these cases when they argue in favor of government intervention. Nevertheless, when one focuses on the contracting process required to attain interoperability instead of market outcomes, it becomes clearer that feasible forms of government intervention are unlikely to address the basic contracting problems more effectively than private ordering.

gains a self-perpetuating market advantage. Because consumers prefer technologies that offer large networks, they tend to favor the dominant firm's technology, even if it is technically inferior. This advantage increases as the relative size of the network increases.[15] The dominant firm may then maintain incompatibilities with its rivals' products to retain control over its installed base, increase the difficulty of entry, and reduce price competition.[16] In addition, vertically integrated dominant firms (i.e., firms that produce more than one system component) have incentives under certain circumstances to foreclose rival producers of complementary components (Ordover, Sykes, & Willig, 1985).[17]

In this way, the exercise of market power can adversely affect interoperability. As happened in the mainframe computer market in the 1970s, users may be isolated in smaller networks, which lowers the value of network technologies. Foreclosure can also reduce competition in markets for complementary services, which can increase prices, reduce consumer choice, and retard innovation.[18] In such cases, policymakers may determine that the public interest requires regulatory intervention to expand the scope of interoperability. For example, the Federal Communications Commission (FCC) has long promoted the benefits of more open and diverse access to the public telephone network. Nevertheless, AT&T and the RBOCs that succeeded it have often resisted efforts to allow rivals access to important functionalities. Thus, in the 1970s, FCC action was required to promote the interoperability of competitively manufactured terminal equipment; in the 1980s, the FCC promoted the concept of Open Network Architecture (ONA) in an effort to facilitate the interoperability of information services and the public telephone network.

Not surprisingly, the tasks that regulators face in their efforts to promote interoperability depend critically on the underlying transactions. In rough terms, regulatory decisions can be divided into easy and hard cases, depending on their characteristics.

---

15. Examples of such network effects are the dominance of VHS video over Sony's Beta system and the dominance of Microsoft's Windows computer operating system.

16. These incentives to maintain incompatibility are, in a way, due to a contracting failure. In theory the dominant firm could craft an enforceable agreement with its rivals to divide the added profits generated from increased demand due to interoperability. Such a contract, however, clearly violates antitrust laws even if it is efficiency-enhancing.

17. There are two conditions necessary for such foreclosure to be in a firm's self-interest. The first condition is that the firm must dominate one stage of production; otherwise, producers of complementary goods could rely on existing competitors or new entrants. The second is that the firm must be constrained in its ability to earn the full monopoly profits from its dominant position in the primary market. Absent such constraints the dominant firm could maximize profits without leveraging its market power, so has no incentives to do so.

18. Economic models provide different predictions on the effect of compatibility on competition. Economides (1991) presents a model in which compatibility leads to decreased price competition. He finds that firms that produce systems composed of two components are able to set higher prices for compatible components because the demand for a single, compatible component is less elastic than the demand for an entire, incompatible system. David and Greenstein (1990), however, review models in which compatibility increases price competition between duopolists.

Regulators face easy cases when transactions involve low levels of uncertainty and asset specificity. Under such conditions, regulatory intervention is unnecessary, or clearly defined and effective interventions exist. In contrast, hard cases arise when high uncertainty and asset specificity are present. Here, all feasible governance forms involve high transaction costs, and policymakers are forced to choose between acutely imperfect alternatives. The policy and political dilemmas that arise out of this choice are examined and then illustrated in two instances of complex cases.

Easy Cases

Cases characterized by low uncertainty and asset specificity do not pose difficult policy problems because rivals or regulators can thwart the efforts of dominant firms to maintain incompatibilities. When a technology is stable and well understood, rival firms that wish to produce compatible system components can do so through reverse engineering or by designing around the dominant firm's technology. Intellectual property laws constrain these strategies, but the trend is moving toward granting fewer protections for interfaces and tolerance of reverse engineering.

An example of this strategy is the rise of the IBM-compatible PC market (Chang, 1994). When IBM introduced the PC in 1981, the computer had an open architecture that facilitated the independent design of peripheral devices and applications software. Nevertheless, IBM sought to prevent competition from compatible personal computers by copyrighting its BIOS—the system that coordinated the communication among the operating system, keyboard, and monitor. Soon after the introduction, however, Compaq and other firms successfully reverse engineered IBM's BIOS, and were able to introduce compatible clones in 1983.

Even when technological complexity is low, a dominant firm may impede rivals if it can refuse to sell complementary components. For example, until the late 1960s, AT&T prohibited its customers from connecting any competitively manufactured terminal equipment, such as answering machines, to its network. Thus, although competing manufacturers understood local loop technology and were able to design fully interoperable pieces of equipment, they could not market them.

However, regulators or antitrust officials can thwart such tactics. When technological uncertainty is low, regulators are able to define interface standards and mandate interconnection. This is essentially what the FCC did in 1975 with its equipment registration program. The commission promulgated a detailed set of standards for the interface between the telephone network and terminal equipment, and ordered AT&T to connect any piece of terminal equipment that abided by these standards.

Hard Cases

Hard cases arise when transactions are characterized by high uncertainty and asset specificity. As argued in the previous section, there are large transaction cost savings from integrating transactions within a firm under such circumstances. Centralized management can mitigate the hazards of opportunism and facilitate the coordination required to adapt to rapidly changing circumstances. However, higher levels of horizontal concentration and vertical integration can also increase firms' market power. Consequently, increased concentration can also degrade market performance by decreasing the competitive pressures for price reductions and innovation, and reducing firms' incentives for interoperability.

A firm's ability to exercise market power critically depends on the presence of uncertainty and asset specificity. For simple transactions, efforts to maintain incompatibilities can be thwarted either by autonomous adaptation by rivals or by regulation. In contrast, when achieving interoperability involves transactional complexity, a dominant firm has access to a powerful, yet subtle, strategy to foreclosure rivals: the refusal to cooperate. When significant technical uncertainties are present, the development of interoperability requires active coordination. Moreover, when the presence of relation-specific investments and frequent changes to market and technological conditions create opportunities for strategic behavior, firms require contractual safeguards to protect investments. However, a dominant firm can simply refuse to extend the cooperation required to address these contractual hazards, thereby imposing significant transaction costs on its rivals. In this manner, the dominant firm can gain market power by raising rivals' costs (Krattenmaker & Salop, 1986). Moreover, this threat is wholly credible because the dominant firm incurs no added costs by such actions.

Thus, cases that combine market power and complex transactions pose a particularly tough policy trade-off. By integrating several system components, a dominant firm can coordinate and promote interoperability, but, at the same time, it has incentives and the capability to maintain incompatibilities with competitors' products.[19] Moreover, benefits of expanding the scope of interoperability and transaction cost savings are both highly uncertain, especially in a dynamic environment. Consequently, evaluating this trade-off is a particularly nettlesome analytical task.

Beyond these analytical difficulties, interoperability poses significant political problems when market power is combined with complex transactions. The dilemma is that all feasible governance structures force some actors to operate under suboptimal contracting arrangements. In an unregulated market, rivals to the dominant firm face significant contractual hurdles because the dominant firm has no incentives to extend the cooperation necessary for rivals to maintain compatibility.

If regulators or antitrust officials intervene, however, they have a limited capacity to address existing contracting failures. When transactional complexity prevents cooperating private actors from crafting adequate contractual agreements, regulators acting with less information, less expertise, and greater political and

---

19. This problem is an example of a more general tradeoff between institutional efficiency and market power that arises in all business mergers (Williamson, 1968).

procedural constraints than private-sector actors are unlikely to perform better. The FCC's ONA policy is a prominent example. The FCC originally envisioned that the ONA rules would be so comprehensive that they would provide independent firms with access to the local telephone network that was equivalent to the access local carriers provided their internal information service operations. However, the FCC has quietly abandoned this lofty goal because it became evident that it could not adequately specify all of the elements of such a contract.

A third policy option is to force the dominant firm to divest its operations in complementary markets. The Department of Justice employed this tactic in its 1982 consent decree which required AT&T to divest its monopoly local operating companies. Although this response successfully removes the unequal contracting status of rivals, it does so by creating new contracting impediments on the dominant firm because it prohibits the firm from undertaking activities it may be well situated to coordinate.

Given that contracting impediments exist under each alternative, some firms will necessarily incur high transaction costs, and frequent disputes are inevitable. Rivals to a dominant firm will complain that it unfairly manipulates its interfaces to their disadvantage, as did IBM's competitors in the mainframe market in the 1970s (Adams & Brock, 1982). Conversely, if regulators place stringent controls on the operations of a dominant firm, it will complain that the restrictions unnecessarily impede its ability to reduce costs and innovate, as have the RBOCs since divestiture.

This combination of factors places policymakers in an untenable political position. They cannot easily ignore these complaints because they are based on real contracting failures; the aggrieved firms can compellingly attest to the innovative services they would provide if freed from existing constraints.[20] Nevertheless, as long as transactional complexity remains high, policymakers have no stable solution. Policy reform only exchanges one set of contracting failures for another, and pressure for reform inevitably resurfaces.

Examples of Hard Cases

To illustrate the policy and political dilemmas that arise in these hard cases, two recent scenarios involving software interoperability in the computer and telecommunications industries are described: (a) the antitrust suit against Microsoft, and (b) the FCC's efforts to assure third parties encess to the advanced network control systems being designed and implemented by local exchange carriers.

MICROSOFT. Microsoft's position in the software market for IBM-compatible personal computers has all the characteristics of a hard case. Because it is the domi-

---

20. In contrast, regulators can more easily identify and discount claims made by firms that are purely strategic uses of the regulatory process to impede competitors.

nant supplier of systems software (e.g., MS-DOS and Windows) and is vertically integrated into markets for applications programs, it may have incentives to impede rivals from marketing compatible software. Moreover, the development of systems and application software involves high asset specificity, in that software designed for one operating system cannot run on other systems without significant added investments. In addition, it involves a high degree of uncertainty. Software developers rapidly create increasingly complex systems that have difficult-to-predict interactions with complementary software components.

Microsoft has been the subject of a long and convoluted antitrust investigation. In 1990, the Federal Trade Commission (FTC) began informally investigating Microsoft. A major issue was whether Microsoft used its control of MS-DOS and Windows to disadvantage rival software vendors.[21] For example, one allegation stipulated that Microsoft had intentionally manipulated Windows to make it incompatible with DR-DOS, a rival operating system. After more than 2 years of investigation, the FTC staff recommended that formal action be taken, but the commissioners twice deadlocked 2-2 on votes, ending the investigation.

The Department of Justice (DOJ) then picked up the investigation, goaded by a senator whose state headquarters a major Microsoft competitor. In July of 1994 after a year-long investigation, the DOJ negotiated a consent decree with Microsoft that narrowly focused on Microsoft's marketing practices and conspicuously ignored the potential anticompetitive effects of intentionally engineered incompatibilities investigated by the FTC. However, this decree was threatened under Tunney Act court review to ensure that the decree was in the public interest. Supported by an unusual amicus brief filed anonymously by several Microsoft competitors, Judge Sporkin, the reviewing judge, vacated the decree, finding that it failed to address consequential anticompetitive practices. The DOJ and Microsoft jointly appealed this decision, however, and the Court of Appeals reversed it, ending antitrust actions after 5 years.

A compelling explanation for the convoluted course of this antitrust investigation is that it involved the political and policy dilemmas inherent in a hard case. Closure was difficult to attain because Microsoft's competitors continually pressured decision makers for action against Microsoft—first in front of the FTC, then the Senate after the deadlock at the FTC, and finally in front of Judge Sporkin.

---

21. The investigation also examined possible efforts by Microsoft to monopolize the market for computer operating systems. The effect of Microsoft's near monopoly on operating systems, however, has a more ambiguous affect on interoperability and social welfare. Because of positive network externalities, wide distribution of a single operating system has significant benefits for computer users. Moreover markets with positive demand externalities tend to tip toward a single standard. Thus, it is uncertain whether Microsoft's marketing practices or natural market processes led to its monopoly position. Because the charge that Microsoft employed incompatibilities in its software is of most interest from the perspective of interoperability, we shall focus on it here.

Given that existing conditions enabled Microsoft to disadvantage its rivals, decisionmakers could not easily ignore these complaints.[22]

The inaction of the FTC and DOJ concerning possible intentional software incompatibilities is equally understandable from a contracting perspective. Assuming that the charges against Microsoft are justified, several factors mitigated against more aggressive action. First, there are large potential benefits from a single firm coordinating changes in systems and application software. Microsoft has developed several technologies that are systemic in nature, including its new Windows 95 operating system and object linking and embedding (OLE), which controls the sharing of information among applications and between applications and the operating system. The value of investments in operating system functionality depends on simultaneous and coordinated investments in application software that effectively employ these functionalities. Moreover, given that these technologies involved a great deal of both technological and market uncertainty, integrated control of design, development, and marketing can lead to important efficiencies.

Second, the Microsoft case remains difficult to adjudicate. Proving in a court of law that Microsoft knowingly engaged in anticompetitive restraints of trade is greatly complicated because the same actions that can have substantial anticompetitive effects in a case where interoperability is important are perfectly acceptable in other circumstances. Protecting trade secrets, refusing to cooperate with competitors, and introducing advances in technology are the normal actions of firms in a competitive market, yet these actions can be anticompetitive when rival firms require close cooperation with the dominant firm to maintain the interoperability essential to compete.

Finally, the FTC and DOJ lacked any compelling mechanism to address the problem of intentional incompatibilities. Feasible remedies would either fail to constrain anticompetitive behavior or constrain this behavior at the cost of reducing the efficiency of existing contractual relations. At one point, the FTC staff did entertain the idea of breaking up Microsoft into separate firms: one to design operating systems and the other to design applications. However, given the possible costs of such action, decisionmakers never seriously considered it. Another option would be to impose a set of rules on Microsoft, requiring the timely release of information and some level of responsiveness to rivals. Given the fluid nature of the software market, however, such rules would be difficult to craft and enforce.

In summary, the results of this lengthy antitrust action have been meager. Microsoft has remained intact, preserving its ability to coordinate the development of systems and application software. Yet the fundamental problem remains: Microsoft's rivals are exposed to inefficiently high contracting hazards. Consequent-

---

22. It may be argued that Microsoft's rivals knowingly did not have any legitimate grievances but abused the process to hamper a powerful rival. It is likely that there is some truth to this interpretation. Nevertheless, one cannot ignore that the preconditions for anticompetitive behavior did exist and that there is at least plausible evidence that Microsoft may have engaged in such practices.

ly, as long as Microsoft retains its dominant position in the operating system market, these firms are likely to continue pressuring policymakers for reform.

ADVANCED INTELLIGENT NETWORKS. The advanced intelligent network (AIN) is a new telecommunications architecture for the software that controls the operation of the public switched telephone network. This architecture will move call-processing software that has traditionally resided in the local switch to centralized service control points connected to switches through data links. The main goals of AIN are to increase the flexibility of network control, and facilitate the rapid development and deployment of customized telecommunications services.

The features of AIN and Microsoft cases are similar. Because it is a radical and complicated departure from previous network architectures, AIN's development involves highly complex transactions. Moreover, AIN is primarily being implemented by the dominant local telephone carriers, the seven RBOCs, and Bellcore, the RBOCs' jointly owned R&D arm. Thus, the development of AIN combines market power with complex transactions. The only significant difference between these cases is that the RBOCs are regulated franchise monopolists, whereas Microsoft is not subjected to entry or price regulation.

AIN raises a number of interoperability issues. The RBOCs have promoted AIN as a way to increase the interoperability of telephone switching equipment. Because the switches designed by different manufacturers are not fully compatible, once a carrier selects its switching vendor, it is bound into a long-term bilateral relationship, and relies on that vendor for software upgrades required to introduce new services. Carriers hope that AIN standards will promote the interoperability of different vendors' equipment and enable them to introduce new services autonomously.

More generally, the manner in which carriers implement AIN will greatly affect the interconnection of competing networks and service providers. Because AIN is not fully standardized, carriers are implementing versions of it based on proprietary elements, raising the possibility that network control functions will not be interoperable across networks.[23] AIN is also likely to have important competitive consequences. For example, independent telecommunications providers' abilities to offer advanced telecommunications services will depend on their access to AIN functionalities. In addition, number portability—a feature considered very important for the promotion of local telephone competition—will rely on AIN capabilities.

The control of AIN by local telephone carriers has raised competitive concerns since its inception. In 1990, a coalition of large telecommunications users and service providers petitioned the FCC to ensure that the RBOCs open AIN to third parties. Their main concern was that the RBOCs would only allow third parties access

---

23. Telecommunications managers, nevertheless, appear confident that carriers will migrate toward standardized versions once they emerge. See *Telephony* May 30, 1994.

to upper level AIN functions, and thus impair the ability of third parties to innovate and compete. It further asserted that the normal standards-setting process would not achieve a desirable level of interoperability because the local carriers lacked incentives to cooperate with potential rivals. Finally, the petitioners requested that the FCC order RBOCs to implement AIN in a modular fashion, with well-defined interfaces between the different levels of software and hardware functionality, and to open access to third parties at all levels.[24]

The FCC took this complaint seriously, recognizing that AIN's architectural barriers could impede competition, and that market forces alone may not provide local telephone carriers with incentives to explore all options of interest to network users. In response, the commission issued a Notice of Inquiry in November 1991, seeking comment on whether greater access to AIN functionalities was warranted. Eighteen months later, it followed with a Notice for Proposed Rulemaking, which proposed serial implementation of AIN access. First, the FCC would mandate access only to high-level functions. Later implementation of access to lower level functionality would be contingent on the gathering of additional information concerning its technical and market implications. The FCC has yet to issue a final ruling.

Similar to the Microsoft case, the FCC's actions reflect the conflict between the recognition that a problem exists and the lack of effective remedies for it. The fact that the FCC opened a rule-making demonstrates the importance of access to lower level AIN functionalities for the interoperability of high-level telecommunications control functions. However, the content of the proposed rule is indicative of the limits of regulatory actions in cases involving highly uncertain technologies. The only certain mandate in the proposed rule—that local carriers provide high-level AIN access to third parties—accomplishes little. This level of access is exactly the type that the original petition complained was inadequate. Moreover, it can hardly be considered an important opening of AIN functionalities because some RBOCs were already planning to offer such access.[25]

The FCC's ability to proceed competently with the serial implementation of more fundamental forms of access remains uncertain. Although Bellcore has issued specifications for the AIN architecture, continued rapid development in key technologies has prevented equipment manufacturers, let alone the FCC, from agreeing on well-defined interfaces for it. Moreover, integrated firms and interfirm alliances continue to dominate the development and marketing of AIN components, indicating the fluid nature of the technology and the importance of close coordination.[26] Even if interfaces are developed, the systemic nature of the AIN

---

24. The petition specifically referred to the widely used Open Systems Interconnection (OSI) seven-layered network model as an example of how to implement appropriate modularity.

25. The willingness of all RBOCs to offer access through service management systems, however, was not certain.

26. Given that a goal of the AIN was to decrease local carriers' dependence on a sole vendor, the success of vendors offering integrated systems solutions is ironic.

means that actors must continue to cooperate closely. For example, when multiple firms control call-routing, a phone call might be issued conflicting instructions. One network's software may try to route the call to one phone number while another network's software wishes to route it elsewhere. Avoiding system breakdown requires the development of message mediation software to resolve these conflicts, as well as cooperation in service testing and problem diagnosis. Consequently, third-party access to fundamental AIN capabilities is likely to remain a problematic regulatory and contracting problem.

The FCC's ability to craft efficient rules for AIN access is further hampered by a lack of information concerning the costs and benefits of AIN interoperability. Local carriers that wish to maintain incompatibilities have strategic incentives to exaggerate the costs of interoperability. At the same time, firms seeking access to AIN hesitate to estimate the benefits of such access because it is difficult to predict the uses of new technologies, and they do not wish to disclose sensitive proprietary information. Consequently, the FCC is in a poor position to craft the type of cooperative agreements required to share the risks and benefits of such investments.

In summary, the FCC will not resolve this regulatory problem easily. If access to lower level AIN functionalities proves to be an important competitive issue, competitors will certainly pressure regulators for relief from perceived constraints on their ability to innovate and compete. Given the coordination problems that arise in the development of such access, however, regulators lack satisfactory policy tools with which to address these problems.

CONCLUSIONS

The interoperability of computer and telecommunications technology poses a novel economic problem that requires a delicate balance between cooperation and competition. In these markets, actors have employed a wide range of governance structures with varying abilities to achieve an appropriate balance. This chapter examined the underlying logic of these structures in contracting terms. Specifically, it analyzed alternative governance structures (e.g., markets, industry standards-setting bodies, joint ventures, integrated firms) as efforts to minimize the transaction costs of achieving interoperability, subject to the characteristics of transactions involved.

This analysis provides both good and bad news for policymakers. The good news is that, under a wide range of circumstances, forms of private governance that arise naturally in competitive markets effectively resolve coordination problems. Policymakers need only permit actors to establish such structures. The bad news is that the difficult cases are particularly intractable. When market power accompanies complex transactions, all feasible regulatory responses place contractual burdens on some actors. Consequently, disputes cannot be avoided, and continual pressures for policy reform are likely.

It is important for policymakers to recognize the nature of this dilemma because failure to do so poses the risk of policy churn, frequent but unfruitful changes in regulatory regimes. Because in hard cases all feasible organizational responses involve contracting impediments, policy reform simply replaces old contracting problems with new ones, creating new pressure for policy reform. The last 30 years of regulation of dominant telephone companies in competitive information services markets are illustrative. Policy has cycled between excluding dominant carriers from these markets and including them subject to various regulatory rules. These frequent changes in policy have been expensive in terms of increased business uncertainty, but regulators have failed to transcend the fundamental dilemma.

The dangers of such policy churn are especially great in the computer and telecommunications industries. Achieving interoperability will continue to involve complex transactions, as long as rapid technological advances continue to press the limits of scientific knowledge and create new markets. In addition, the dominance of certain network components by a single firm will be a continuing concern. The advantages of current monopolies will only whither slowly, and because these markets are characterized by network externalities, new dominant firms will arise when markets standardize on a firm's proprietary technology.

Policymakers do not lack tools for addressing these problems. Well-specified standards reduce transactional complexity, and hence limit the need for integrated governance and the ability of dominant firms to disadvantage rivals. Thus, policies that facilitate standards setting can reduce the prevalence of hard cases. Governments can accomplish this through financial support of the industry standards-setting process, through its own purchasing policies, and through its support of R&D.[27] In addition, increased competition reduces firms' incentives to maintain incompatibilities. Thus, a strong policy promoting competition at all levels of a systems technology is warranted.

Finally, policymakers should be patient. The corrosive combination of market power and complex transactions is likely to arise in new segments of the telecommunications and computer industries. Nevertheless, the pernicious trade-offs related to a specific system component are likely to fade with time. As technologies mature, technical and market uncertainties subside, thereby reducing transactional complexities and facilitating competition. In addition, old technologies face competition from new ones, thereby limiting the incentives to maintain incompatibilities. For example, as the rise of the personal computer has eclipsed IBM's preeminence in the computer industry, concern over its dominance of the mainframe market has subsided. Similarly, future technological advances are likely to eclipse both Microsoft and local telephone carriers. Given the limitations of feasible policy interventions aimed to

---

27. Weiss and Spring (1994), for example, suggest that alternative mechanisms for financing the standards process such as direct governmental subsidies can improve the standards-setting process.

promote interoperability, patient reliance on these long-run market processes is warranted.

REFERENCES

Adams, W., & Brock, G. (1982). Integrated monopoly and market power: System selling, compatibility standards, and market control. *Quarterly Review of Economics and Business,* 22(4), 29–42.

Besen, S., & Johnson, L. (1986). Compatibility standards, competition, and innovation in the broadcasting industry (R-3453-NSF). Santa Monica, CA: RAND.

Chang, I.Y. (1994). *The economics of dominant technical architectures: The case of the personal computer industry* (P-7888). Santa Monica, CA: RAND.

Computer Systems Policy Project. (1994). Perspectives on the national information infrastructure: Ensuring interoperability. http://www.cspp.org/reports/.

David, P. A., & Greenstein, S. (1990). The economics of compatibility standards: An introduction to recent research. *Economics of Innovation and New Technology, 1,* 3–41.

Economides, N. (1991). Compatibility and the creation of shared networks. In M.E. Guerin-Calvert & S.S. Wildman (Eds.), *Electronic service networks: A business and public policy challenge.* New York: Praeger, 39–55.

Economides, N., & White, L. (1993). *One-way networks, two-way networks, compatibility, and antitrust.* New York University Department of Economics Working Paper Series EC-93-14.

Farrell, J., & Saloner, G. (1985). Standardization, compatibility, and innovation. *Rand Journal of Economics, 19*(2), 70–83.

Farrell, J., & Saloner, G. (1988). Coordination through committees and markets. *RAND Journal of Economics, 19*(2), 235–254.

Garud, R., & Kumaraswamy, A. (1993). Changing competitive dynamics in network industries: An exploration of sun microsystems' open systems strategy. *Strategic Management Journal, 14,* 351–369.

Greenstein, S. (1990). Creating economic advantage by setting compatibility standards: Can "physical tie-ins" extend monopoly power? *Economics of Innovation and New Technology, 1*(1/2).

Katz, M., & Shapiro, C. (1985). Network externalities, competition, and compatibility. *American Economic Review, 75*(3), 424–440.

Knight, F. (1965). *Risk, uncertainty, and profit.* New York: Harper & Row.

Krattenmaker, T., & Salop, S. (1986). Anticompetitive exclusion: Raising rivals' costs to achieve power over price. *The Yale Law Journal, 96*(2), 209–293.

Ordover, J., Sykes, A.O., & Willig, R. (1985). Non-price anticompetitive behavior by dominant firms toward the producers of complementary products. In F. Fisher (Ed.), *Antitrust and regulation: Essays in memory of John McGowan.* Cambridge, MA: MIT Press, 115–130.

Sirbu, M., & Weiss, M. (1990). Technological choice in voluntary standards committees: An empirical analysis. *Economics of Innovation and New Technologies, 1*(1/2).

Teece, D. (1988). Technological change and the nature of the firm. In G. Dosi et al. (Eds.), *Technical change and economic theory*. London: Pinter Publishers.

Weiss, M.B.H., & Spring, M.B. (1994). *Alternatives to financing the standards development process*. Paper presented to the 22nd annual Telecommunications Policy Research Conference, October, Solomons, MD.

Williamson, O.E. (1968). Economics as an antitrust defense. *American Economic Review, 58*, 18–35.

Williamson, O.E. (1975). *Markets and hierarchies: Analysis and antitrust implications*. New York: The Free Press.

Williamson, O.E. (1985). *The economic institutions of capitalism*. New York: The Free Press.

Williamson, O.E. (1991). Comparative economic organization: The analysis of discrete structural alternative. *Administrative Science Quarterly, 36*, 269–296.

10

Comparing Integrated Broadband Architectures From an Economic and Public Policy Perspective

Nosa Omoigui
Microsoft Corporation

Marvin A. Sirbu
Carnegie Mellon University

Charles Eldering
General Instrument Corporation

Nageen Himayat
General Instrument Corporation

Current residential telecommunications services are limited to telephony in the form of Plain Old Telephony Service (POTS), narrowband Integrated Services Digital Network (ISDN), and broadcast analog video services. In the near term, video conferencing will create a demand for telecommunications services at rates of n × 64 kb/s, or using new virtual circuit technologies such as asynchronous transfer mode (ATM). Future video services will include broadcast digital video and video–dial-tone services.

Of all the various architectures that have been considered for provision of such services, Hybrid Fiber-Coax (HFC) and Fiber-to-the-Curb (FTTC) are the main contenders. HFC is popular in the cable industry, and is currently used to provide analog broadcast services. It will also be used to provide digital broadcast and interactive services in the future. The FTTC system is an embellishment of the Digital Loop Carrier (DLC) architecture, which traditionally was used by Regional Bell Operating Companies (RBOCs) to provide telephony services. In the FTTC system, an Optical Network Unit (ONU) is used to serve a small number of subscribers (typically between 8–32) using a fiber connection from the central office. FTTC is being seriously considered by the telephone companies.

DESCRIPTION OF THE BASIC ARCHITECTURES

This section describes the different architectures being considered for the provision of broadband interactive services. These architectures can be further categorized into two levels of sophistication: the plain vanilla flavors, which provide only telephony or analog video services, and the upgrades to these same architectures, which are designed to provide interactive video and other advanced services.

Architectures for Narrowband or Broadcast Services

This section focuses on three baseline architectures designed for telephony or broadcast video service provision: HFC, DLC, and FTTC.

HFC ARCHITECTURE. The HFC architecture, as the name implies, employs a combination of both fiber and coaxial cable. HFC employs a fiber backbone that corresponds to the fiber feeder in the DLC and FTTC networks. HFC can be used to deliver both distributed video and telephone services, and can be made to carry up to 500 channels (for a 500-home node, this implies one channel per home). In addition, HFC systems can simultaneously support Video on Demand (VOD) and telephony. For broadcast transmission, HFC provides each subscriber with the same group of video channels—therefore, there is high sharing of resources. The HFC head-end receives signals from local studios, over-the-air broadcasts, or microwave and satellite sources, and then combines and retransmits these signals over the trunk (or feeder) cable of the network. Portions of the signal are then split to feeder cables and, in turn, to drop cables to serve the household. The HFC distribution plant consists mainly of coaxial cable. Because of the large attenuation of signals over coaxial cable, feeder amplifiers (minibridgers and line extenders) are necessary to amplify the signal for both forward and return paths. Passives include taps, connectors, three-way splitters, power inserters, and directional couplers. Sixty-volt power supplies supply power needed to drive the actives in the distribution network. The drop loop consists of coaxial drop cable. An addressable converter is needed on the subscriber premises to deliver distributed and switched video services.

The HFC system is a shared transport medium; signals for all the subscribers fed from a given node are available at every household served by the node. This requires encryption to prevent unwarranted reception by some subscribers. In the return path, signals from all households are summed at the node and transmitted to the head-end. The fact that all signals from the subscriber locations are received at the head-end results in the phenomenon called *noise funneling*, in which un-

Figure 10.1
Diagram showing main components of the HFC architecture

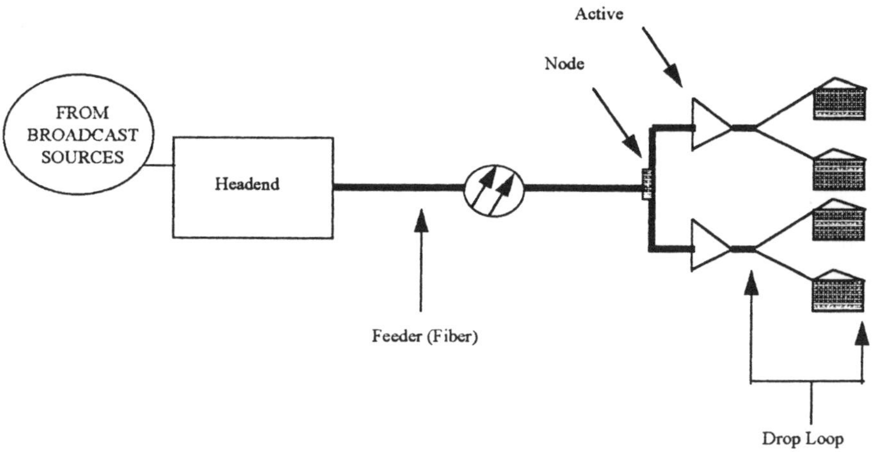

wanted signals (noise and ingress) from all the subscriber locations appear at the head-end. This requires the use of a multiple access technique—such as Time Division Multiple Access (TDMA), Frequency Division Multiple Access (FDMA), or Code Division Multiple Access (CDMA)—to allow the various subscribers to transmit to the head-end. The use of fiber-optic transmission systems from the head-end to the node has resulted in a decrease in node size, with the typical node size today being in the range of 500–2000 homes passed. At node sizes of 500, the ingress problems are greatly reduced, and the amount of spectrum available in the downstream and upstream would support narrowband services at full penetration rates, as well as interactive video services. Figure 10.1 shows the main components of an HFC architecture.

DLC ARCHITECTURE. Digital Loop Carrier (DLC) is a fiber-based architecture that is typically used by telephone companies to provide narrowband telephony service (Plain Old Telephone Service (POTS); Omoigui, Sirbu, Eldering, & Himayat, in press). The DLC system analyzed for this chapter has terminals both in the Central Office Terminal (COT) and Remote Terminal (RT). The system employs the Synchronous Optical Network (SONET) signal format as the transmission medium between terminals. Each terminal can support two independent directions of SONET transmission, and consists of three major subsystems: a common control assembly (CCA; optical receive and transmit), modulation and demodulation of SONET carrier, and SONET formatting. Each CCA can support up to nine channel bank assemblies at the OC-3 rate, each of which supports up to 56 channel cards. Each channel

Figure 10.2
Diagram showing main components of the DLC architecture

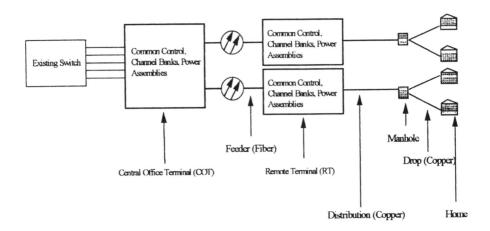

card can support up to four lines. Therefore each terminal can support up to 2,016 lines. The channel bank assembly houses the various channel units and channel bank common equipment units. The DLC distribution plant is composed of twisted pair copper cable of different sizes, with appropriate connectors, splices, and pedestals. The drop loop includes pedestals at the curb and copper drop cable. Subscriber premises equipment includes a protective block for surge protection. Figure 10.2 shows the main components of a DLC architecture.

FTTC ARCHITECTURE. The FTTC architecture is similar to the DLC system. In an FTTC system, fiber is extended past the node and to the pedestal, which is at the curb near the home (Omoigui et al., in press). This has several advantages. First, it exploits the high capacity of fiber by sharing the transmission facilities over more subscribers in the distribution network Second, by deploying more fiber, it lowers the incremental investment needed for broadband services. The drop from the curb is assumed to be twisted copper pair, although if FTTC is to be used to provide video, a dual drop cable containing both telephony twisted pair and coax may be cost-effective in comparison with the cost of the electronics needed to multiplex and demultiplex telephony and video bandwidth signals over a twisted copper pair drop alone. In an FTTC system, remote electronics—referred to as a Host Digital Terminal (HDT)—takes a high bit-rate signal from the COT and demultiplexes it via a fiber bank assembly onto lower bit-rate distribution fibers. The HDT may also contain a Network Power Assembly (NPA) to provide power to the Optical Network Units

Figure 10.3
Diagram showing main components of the FTTC architecture

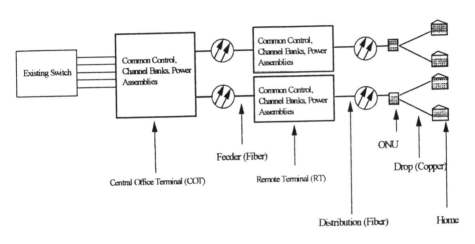

(ONU) located at the curb. When power comes from an NPA, the distribution fiber cable must include a copper power feed. Fiber bank assemblies provide the optical transmit, receive, and multiplexing functions for the optical link to the ONUs. The ONU contains the channel units or line cards that interface a customers copper wire-based equipment to the terminal. The FTTC distribution plant is composed of fiber cable of different sizes, again including splicing and connector costs. The drop loop and subscriber premises equipment are essentially the same as in DLC.

This narrowband architecture is well suited for reliable delivery of POTS, IS-DN, and special services, and is well positioned to meet the higher bit-rate demands for remote LAN access and video–telephony applications. A key difference between the HFC and FTTC architectures is the means of transporting digital signals from the RT to the curb. FTTC uses baseband transport of digital signals over low-noise fiber; HFC requires the modulation of a Radio Frequency (RF) carrier with the digital signal for carriage over coax from the remote node. Both systems use copper for the drop. In an FTTC system, copper twisted pair for the drop creates more noise exposure than coax drops, which are the norm in an HFC system. The main components of an FTTC architecture are shown in Figure 10.3.

Architectures for Broadband and Interactive Services

Modifications to the baseline infrastructures described earlier are required if these architectures are to support the full range of services encompassing analog broadcast, interactive video, and telecommunications services. These extended architectures are now described.

ADVANCED HFC. One approach to these advanced networks is to upgrade the basic HFC infrastructure. An Advanced HFC (AHFC) architecture [2] is shown in Figure 10.4. Additional equipment is now required at the head-end to support interactive video and telecommunication services. The HDT must be capable of interfacing to an ATM network as well as a Public Switched Telephone Network (PSTN) to accept video and telecommunications traffic. The digital data streams are then combined with broadcast signals and switched to individual subscribers requiring service. Subscribers communicate with the head-end via set-top boxes for low bit-rate interactivity suitable for video–dial-tone applications. Return telecommunications traffic is coupled to the return channel through a premises interface device termed the *coaxial termination unit*, which is placed on the side of the residence.

From a technical perspective, the challenges to this approach include managing signal security via access control and encryption, as well as overcoming impairments on the return channel for providing highly reliable service required by telecommunications applications. However, there is no fundamental reason that the HFC network cannot be upgraded to provide the full range of telecommunications services. As is shown, the HFC network is an extremely cost-effective transport medium.

Figure 10.4
Diagram showing main components of the advanced HFC architecture

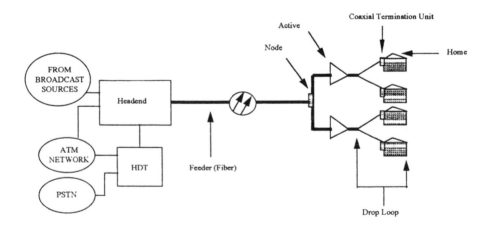

FTTC WITH HFC OVERLAY. Assuming a telephone company plans on installing an FTTC network to support telephony and narrowband switched services, there are two alternative routes for providing broadcast services: using an overlay HFC network, or upgrading the FTTC network to carry video [2]. Typically, an HFC network will carry 30–100 channels simultaneously, in analog form. This allows the recipient to use todays cable–ready TV sets to receive the signal. If the HFC system is installed at the same time as the FTTC system, the net additional costs of the overlay HFC are quite small because the two systems can be pulled simultaneously. A siamese drop cable, with both coax for HFC and twisted pair for FT-TC, can be pulled for the cost of pulling either one singly. By contrast, providing broadcast services over FTTC is more costly.

Indeed, it is not economical to provide video services in a broadcast mode over FTTC. Wideband lasers capable of carrying broadband signals are expensive. The alternative of providing digital transport and digital-to-analog conversion at the ONU for up to 100 broadcast channels is also prohibitively expensive. Instead, the FTTC network must be upgraded with digital ATM switches to provide a video–dial-tone capability that allows the customer to dial up either broadcast services or VOD. In addition, because FTTC systems carry video in digital form, not analog, the consumer needs a high-cost (~$350) set-top box to decompress the video signal and convert it to a form suitable for use with analog TV sets. Once

Figure 10.5
Diagram showing main components of the FTTC
with HFC overlay architecture

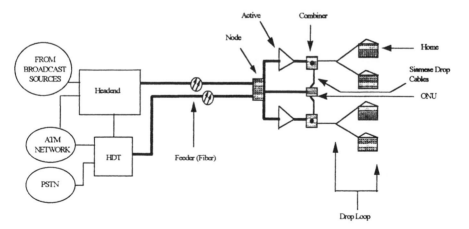

these upgrades have been made, however, the FTTC system can provide interactive video services (IVS), or switched video services (SVS), such as VOD, for no additional cost. By contrast, if the analog HFC overlay route was chosen, additional investment would be required to upgrade either the HFC or the FTTC to carry IVS. Three options are considered starting with FTTC for telephony: (a) FTTC for telephony and HFC for both broadcast and IVS services (FTTC[T]-HFC); (b) FTTC for telephony, HFC for broadcast, and FTTC for IVS services(FTTC-HFC[B]); and (c) FTTC for telephony, broadcast, and IVS services (i.e., all FTTC—broadcast provided identically to IVS). The FTTC–HFC hybrid architecture is shown in Figure 10.5. The FTTC–HFC convergence is accomplished close to the living unit in a combining tap that serves the premises with the necessary physical media.

ALL FTTC. As noted previously, an FTTC network could provide switched digital access to today's broadcast and premium analog cable channels, as well as to VOD. Switched digital access means that the consumer would be obliged to have a set-top box that ordered the appropriate channel to be delivered down the line, and converted the signal from digital to analog form. However, the Cable Act of 1992 directed the Federal Communications Commission (FCC) to issue rules designed to oblige cable service providers to support the provision of multiplexed analog signals in a form compatible with todays cable-ready TV sets.[1] The required interposition of a set-top box in an FTTC-only system may require a waiver from these rules if the FTTC system is construed as providing "cable service." Telephone companies desirous of providing "cable service," but wary of seeking a waiver under the current rules, may believe themselves obligated under the rules to construct an HFC overlay on top of an FTTC system, rather than service all video needs through a single FTTC network as switched digital video, regardless of which method is less expensive. Because of these rules the all FTTC option is analyzed here to shed light on the actual costs to the telephone companies and consumers, should the FCC refuse to grant the necessary waiver. In the event that waivers from the Cable Act mandate are available, the all-FTTC option must be considered on its economic merits.

DLC WITH HFC OVERLAY. Like FTTC, DLC can also be upgraded to support interactive services by overlaying it with an HFC network. In this scenario, the HFC network carries broadcast services and IVSs, whereas the DLC network carries only telecommunications services. The architecture is similar to that of the FTTC–HFC overlay architecture, where HFC is used for both analog and IVSs. Unlike FTTC, it is not possible to carry video directly on the DLC system at acceptable

---

1.   47 USC 544a and CFR Sec. 76.630.

bit rates.[2] Telephony is provided over the DLC portion of the network, which uses copper in the distribution plant and eliminates the ONUs of FTTC.

ECONOMIC MODELS

This section presents an economic comparison of HFC- and FTTC-based architectures when they are deployed to carry a full range of services. In particular, cost[3] analysis of the HFC architecture supporting analog broadcast, interactive video, and telecommunications services is presented.

Scenarios Considered

The economic models developed in this study considered the following scenarios for the deployment of telecommunications and interactive video services:

1. HFC network carries broadcast services and IVSs, and DLC network is used for telecommunications services.
2. HFC network carries broadcast and IVSs, and the FTTC network is used for telecommunications services.
3. HFC network carries broadcast services only, and IVSs are placed on the FTTC network in addition to the telecommunications services.[4]
4. All services are carried over the HFC network, in the Advanced HFC configuration.
5. All services are carried over the FTTC network.

Cost comparison of these architectures is based on developing accurate cost models for laying out the HFC and FTTC infrastructures to enable them in their traditional roles of supporting broadcast and telecommunications services, respectively. The costs of providing IVSs over the HFC and the FTTC architectures are determined separately. The incremental costs are then combined with the infrastructure costs to obtain the overall results.

Assumptions

The set of assumptions governing the costs models are outlined next. These assumptions are representative of current visions of network deployment.

---

2. Using Asymmetric Digital Subscriber Loop (ADSL) technology, video could be delivered over a DLC infrastructure, but ADSL will not support bit rates needed for HDTV or real-time compression of action video (e.g., a football game broadcast as it is played). .

3. In this analysis, the term *cost* refers to the total installed cost to a network operator, all at market prices.

4. This hybrid architecture is being actively considered by a number of RBOCs for carriage of analog broadcast services that presently are best supported over an HFC infrastructure.

1. The node size is assumed to be 480 homes, along with a central office size of 30,000 homes and a network head-end size of 180,000 homes.

2. The plant is assumed to be 70% aerial and 30% underground, with a 60% take rate for analog services. For modeling, the cost of the penetration rate for IVSs is set to be a variable assumed to be less than the take rate for analog services.

3. A set-top box estimated to cost $125 is needed in conjunction with an HFC network carrying broadcast analog services. A digital set-top box, which is needed for all IVSs and for enabling the all FTTC network to carry broadcast services, is estimated to cost $350.

4. The peak coincident busy hour usage of these services is considered to be fixed at 25% of the interactive service penetration rate. In addition, it is assumed that no further cabling infrastructure resources are required to support IVS.

5. IVS offerings are considered to be 70% requiring 1.5 MB/s and 30% requiring 6Mb/s, with a ratio of 1.6 set tops per user. This corresponds to an average bandwidth of 2.85 Mb/s per IVS channel. This imposes a relatively low demand on bandwidth per channel. Demands on system bandwidth may be higher, in general, and would certainly be higher if HDTV services are to be introduced.

For modeling the cost of SDV services over the FTTC architecture, the system proposed by Broadband Technologies (BBT) is used as a reference. IVSs over the FTTC architecture supports very high bandwidth in the forward direction (approximately 2.4 Gb/s, or an equivalent of 400 channels at 6Mb/s is provided per node). Channels are switched to individual subscribers via ATM switches at the head-end. Each subscriber is provided the capability of receiving six video channels. High-bandwidth return capability is provided, and the telecommunications traffic is transported in combination with video information.

The FTTC architecture used in this analysis represents a custom development carried out by BBT that attempts to optimize ATM switching technology to account for asymmetric bandwidth requirements of IVS in the forward and return directions. Additional optimization is also carried out by combining, with video, telecommunications data streams already present on the FTTC networks. From the viewpoint of cost modeling, it becomes difficult to assign costs for a custom-developed architecture unless detailed system designs are available. Hence, the approach has been to estimate the cost of a system providing SDV functionality equivalent to that of the telecommunications services. Telecommunications and video services are treated independently, but this assumption is not allowed to affect the cost estimate significantly. This approach is useful in providing a rough estimate of the incremental cost of SDV services over FTTC, and, as is shown later, is extremely useful in comparing the general trends in cost with respect to both architectures. The following assumptions are used in the incremental cost analysis of FTTC:

1. A node size of 1,000 is chosen; the central office and head-end size are assumed the same as for HFC.
2. A 100% penetration rate is assumed for telecommunications services, therefore additional infrastructure equipment is not required for deploying video services.
3. One ONU is deployed for every 16 subscribers, and it is assumed that the IVS subscribers are uniformly distributed amongst all ONU service areas.[5]

In both the HFC and FTTC instances, cost comparison is carried out on a cost-per-home-passed basis. A 40% markup is also applied to the incremental costs for IVSs to account for installation costs. Additionally, for comparing the cost of IVSs over the HFC or FTTC network, a suburban plant (composed of 100 homes per mile passed) is assumed with node sizes being equivalent to those considered for the IVS cost modeling.

Summary of Component Costs and Economic Assumptions

Tables 10.1 and 10.2 provide more detail on the underlying component costs assumed in the economic analyses. Economic models were developed using the analytica engine from Lumina Decision Systems. For certain variables such as trenching costs, which vary with terrain, a probabilistic distribution of cost was assumed, rather than a fixed estimate. The final cost numbers reported herein should thus be understood as the mean value of the resulting probabilistic distribution of total costs. Installation costs are based on a sample of contemporary projects. Component costs are based on several manufacturers' list prices, adjusted for estimated quantity discounts.

TABLE 10.1
Set Top Costs and Penetration Assumptions

Variable	Value
Cost of analog set top	$125
Cost of digital set top	$350
Penetration analog video	60%
Number of set tops per subscriber	1.6

---

5. This implies that, for the most part, the cost of supporting IVS subscribers is fixed, regardless of the penetration rate and the slight variation in cost due to the need for additional line cards for each IVS subscriber.

TABLE 10.2
Estimated Costs of Common Network Components

Component	Cost	Units
Single-mode fiber	0.2 (the number of fibers)	$/meter
Feeder fiber sheath	0.6	$/meter
Feeder inner duct	1.15	$/meter
Fiber splice	normal(34.4, 11.5)[a]	$/splice
Fiber pigtail	250	$/unit
Power per watt (WOW)	10 in C.O., 15 in R.T.	$/watt
Fiber strand installation cost	normal(1108.8, 153.94)	$/mile
Fiber lash installation cost	normal(1963.8, 353.7)	$/mile
Fiber duct installation cost	normal(2227.8, 673.01)	$/mile
Trenching cost	normal(14870, 5833.9)	$/mile

a. This notation means the component cost is assumed to be normally distributed, with a mean of $34.4 and a variance of $11.5.

As noted earlier, in the DLC architecture, an RT consists of a cabinet, channel bank assemblies, common control, and power equipment. The total equipment cost of a fully equipped suburban node serving 1,000 homes is $219,000. Table 10.3 notes some of the key component costs that went into calculating the previous figure, as well as other key cost assumptions for the DLC models.

TABLE 10.3
Table of Estimated Costs of Main DLC Specific Network Components

Location	Component	Equipment cost ($)	Installation cost ($)
Central office (COT), remote node (RT)	Optical transmitter unit (OTU)	1,170	40% markup
Central office (COT), remote node (RT)	Optical receiver unit (ORU)	1,070	40% markup
Central office (COT), remote node (RT)	SONET formatter unit (SFU)	2,520	40% markup
Remote node (RT)	Channel bank assemblies	$90,000/1,000 homes	40% markup
Remote node (RT)	Line card (included in channel bank assemblies)	118/4 lines	40% markup
Distribution	Copper splice	1.66	n/a
Distribution	Copper connector	20	n/a
Distribution, drop	Copper inner duct	0.75 ($/meter)	n/a
Distribution	Copper cable (twisted pairs; varies by cable size)	1.65–10.8 ($/meter)	aerial - normal (3.75, 1.2); underground— normal (4.75, 1.5)

In the FTTC architecture (Table 10.4), fiber-based channel bank assemblies at the Host Digital Terminal (HDT) carry signals to an ONU, from which copper drops extend to each household. The HDT may be located either in the field—similar to a DLC RT—or inside the central office. A fully equipped and installed HDT supporting 1,000 homes costs $283,000.

TABLE 10.4
Table of Estimated Prices of Main FTTC Specific Network Components

Location	Component	Equipment Cost ($)	Installation Cost ($)
Remote Node (HDT)	Fiber Bank Assembly	11,550	40% markup
Optical Network Unit (ONU)	24 line ONU with line cards	1,990	40% markup
Remote Node (HDT)	Network Power Supply (NPS)	473	40% markup
Remote Node (HDT)	Network Power Controller (NPC)	173	40% markup

The costs of the main infrastructure components of an HFC network are shown in Table 10.5.

TABLE 10.5
Table of Estimated Costs of Main HFC Specific Network Components

Location	Component	Units	Equipment cost	Installation cost
Drop loop	RJ-6 drop cable (aerial)	$/ft	0.095 for rural and suburban plants; 0.112 for urban	see below
Drop loop	RJ-6 drop cable (underground)	$/ft	0.084 for rural and suburban plants; 0.117 for urban	see below
Distribution	Directional coupler	$	19.75	25
Distribution	Power inserter	$	19	25
Distribution	3-way splitter	$	25	25
Distribution	Minibridger	$	769	45
Distribution	Line extender	$	189	35
Distribution	2-port tap	$	7.4	15
Distribution	4-port tap	$	8.53	15
Distribution	8-port tap	$	13.28	15
Distribution	60V power supply	$	1,710.28	25
Head-end	Forward laser and cabinet	$	5758	137
Head-end	Return path receiver & chassis	$	795	137

(table continued on next page)

(Table 10.5 continued)

Head-end	Commander modulator	$	1,300	275
Head-end	Commander demodulator	$	3,000	275
Remote node	Forward costs	$	2,191	375
Remote node	Return costs	$	765	
Subscriber premises	Set top	$	125	50
Distribution	750 coaxial cable	$/ft	0.91 for under-ground, 0.752 for aerial	
Distribution	500 coaxial cable	$/ft	0.45 for under-ground, 0.359 for aerial	
Distribution	Aerial coax strand installation	$/mile	normal (1,050.7, 194.38)	
Distribution	Aerial coax standard install cable	$/mile	normal (1,322, 9.08)	
Distribution	Coax trench installation	$/mile	normal (12,950, 4,191.1)	

To deliver digital video over an HFC system, it must be enhanced with additional equipment for carrying wideband digital signals. Typically, a basic carrier rate of 29 Mbps per 6 Mhz video channel is used for carrying video traffic. The number of channels that can be carried on a single 29 Mbps carrier depends on the compression rate: Live action typically requires more bps than a movie, which can be preprocessed. HDTV signals consume four to six times the bit rate of NTSC signals. A digital carrier totaling 29 Mbps representing 4–10 channels will be transmitted by satellite to the cable head-end, where an Integrated Transencryption Multiplexor (ITEM) decrypts the signal and then reencrypts it for transmission over the cable. This digital signal is modulated for transmission over the fiber backbone using a 64-QAM modulator. Return signals from the set top boxes are modulated for transmission upstream from the remote node using a Quadrature Phase-Shift Keying (QPSK) modulator. A network control unit processes the upstream signals from the set tops.

Table 10.6 summarizes the costs of the electronic components of an AHFC network.

For ATM costs, prices have been falling at 30% per year. These prices, which are shown in Table 10.7, are as of early 1995. Also, it is assumed that the cost of an OC-12 port would be four times that of an OC-3 port.

TABLE 10.6
Prices for Advanced HFC Digital Video components

Item	Cost ($)
ITEM	50,000
64-QAM modulator	3,500
QPSK modulator	2,500
QPSK receiver	10,000
Network control unit	30,000

TABLE 10.7
Estimated Costs of ATM Components ($)

Item	Cost
Cost of fore systems ASX200 ATM Switch (2.4 Gbps switch)	21,950
Cost of ATM output port	2,500
Cost of ATM input port	10,000

Table 10.8 shows the cost breakdown of the total per home cost of the DLC and FTTC architectures. These numbers are for a 1,000 home, suburban topology.

TABLE 10.8
Cost Breakdown of Total Cost per Home Passed
(DLC and FTTC)—1,000 Home Node, Suburban Topology

Location/Architecture	DLC ($)	FTTC ($)
Feeder electronics	164.29	146.46
Feeder electronics labor	65.72	58.59
Feeder cable	20.10	20.09
Feeder cable labor	31.24	31.24
Distribution electronics	309.10	304.15
Distribution electronics labor	123.70	121.70
Distribution cable	196.80	151.27
Distribution cable labor	107.83	90.31
Drop electronics	75.00	305.11
Drop electronics labor	91.20	183.20
Drop cable	60.28	54.09
Drop cable labor	117.59	224.06
Total cost	1,362.85	1,690.27

RESULTS AND ECONOMIC COMPARISONS

The results of the analysis can be best understood by dividing the architectures into two categories: baseline architectures, which only provide analog broadcast and telecommunication services, and advanced architectures, which also provide interactive broadband services.

Baseline Architectures

Figure 10.6 shows the baseline costs for HFC, FTTC, and DLC. These are not directly comparable of course, because the baseline HFC system provides only broadcast video, whereas the other two provide only telecommunications. Still, the higher cost for FTTC over DLC to provide basic telephony can be noted. This suggests that FTTC can be cost justified only in terms of future services, such as video.

Advanced Architectures

To provide telephony services, an HFC network must be augmented with additional electronics and premises equipment. These costs have been estimated at $400 per household, with the assumption that 100% are telephone subscribers. Similarly, to provide broadcast video services, an FTTC or DLC network must be augmented by an overlay HFC network. Further investment is needed to bring any of

Figure 10.6
Price per home passed for baseline HFC, DLC, and FTTC networks

(Suburban plant, 1000-home node)

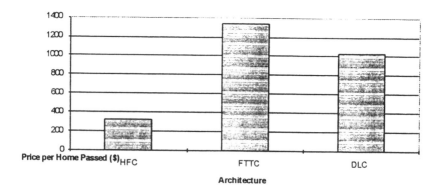

Figure 10.7
Cost comparison of all services over hybrid/single architectures

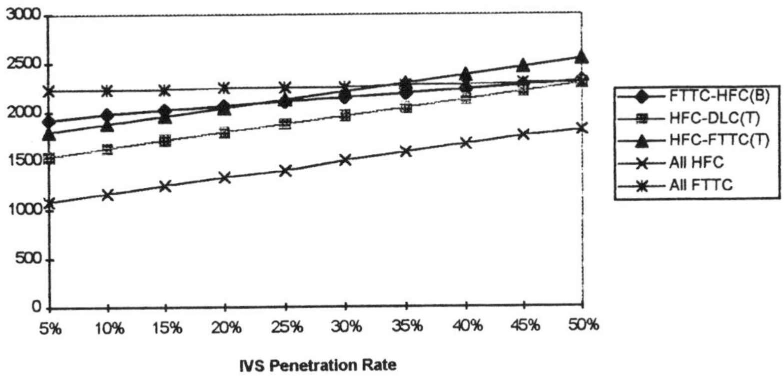

IVS Penetration Rate

Key:
HFC-DLC(T): Broadcast and Switched Digital Video (SDV) services over HFC and telecommunications over DLC (Scenario 1); HFC-FTTC(T): Broadcast and SDV services over HFC and telecommunications over FTTC (Scenario 2); FTTC-HFC(B): Telecommunications and SDV services over FTTC and analog broadcast over HFC (Scenario 3); All HFC: Broadcast, SDV, and telecommunications services over HFC (Scenario 4); and All FTTC: Broadcast, SDV, and telecommunications services over FTTC (Scenario 5)

the three architectures to the point where they can handle Switched Digital Video (SDV) services as well.

Figure 10.7 shows the combined result for the five scenarios outlined in the previous section. It is seen that the most cost-effective solution is to use the HFC architecture to carry all services. This primarily results from the fact that the initial infrastructure costs for HFC are significantly lower when compared with the FTTC/DLC infrastructure costs (Omoigui et al., in press).

Figure 10.7, shows that the AHFC architecture is the least costly at all IVS penetration rates—between 5%–50%. This is followed by the HFC–DLC hybrid architecture, where HFC carries IVSs and DLC carries telecommunication services. Within this same range of penetration rates, the remaining three scenarios have comparable costs. However, as the take rate increases, the cost for the HFC–FTTC hybrid architecture (where FTTC carries telecommunication services) crosses that of the all FTTC and the FTTC–HFC (where HFC carries only analog broadcast services). Indeed, the three lines seem to cross at a penetration rate of 25%. At higher IVS penetration rates, the all FTTC option is preferred.

Figure 10.8
Comparison of the incremental cost of switched digital video (SDV)
services over HFC and FTTC

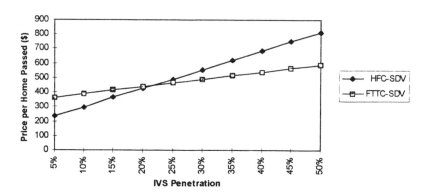

Price per Home Passed (IVS only) - 500 Homes for HFC and 1000
Homes for FTTC

Fig. 10.9
Incremental cost of SDV services over HFC
as a function of bandwidth
(273.6 Mbps corresponds to the bandwidth requirement at 50% IVS penetration rate, 25%
coincidental usage (96 Channels) and an average bandwidth of 2.85 Mbps per channel)

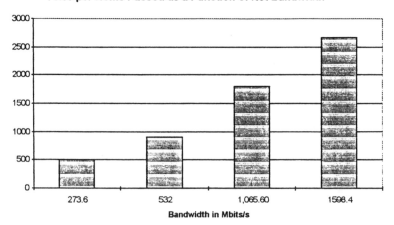

Price per Home Passed as a Function of Net Bandwidth

FTTC costs are essentially flat with the penetration of IVSs, but HFC costs increase steadily with penetration. This is illustrated in Figure 10.8, which shows the incremental cost of providing Switched Digital Video (SDV) services over HFC and FTTC architectures. This also implies that at IVS penetration rates higher than 50%, the all-FTTC option would become significantly cheaper than all the hybrid architectures and would eventually become as cheap as the all-HFC architecture.

It is seen that the IVS cost over HFC increases with increasing penetration rate, whereas IVS cost over FTTC is relatively insensitive to the penetration rate. This can be explained by the fact that the FTTC structure provides for a large bandwidth that is fixed, regardless of the penetration rate. This implies that fixed costs are incurred for the ATM switch and the ATM switch interfaces to the ATM network. Additionally, the position of the ONUs are fixed by the demand for telecommunications services. Therefore, even if a few subscribers are present in an ONU service area, high-bandwidth output ports in the ATM switch and the optical units within the ONU are still needed to support them. The only component cost that changes with the penetration rate is the cost of subscribers video channel cards. Hence, for low penetration rates, IVS over FTTC is more costly than that over HFC, but the cost does not change significantly with increasing penetration rate.

In contrast, the cost of IVS over the HFC network grows as the demand for bandwidth through the system grows. The increasing costs of IVS service over HFC as a function of system bandwidth is shown in Figure 10.9. These figures do not include set top costs, and are shown here primarily to illustrate how incremental costs would rise with increasing bandwidth. The bandwidth of 273 Mbps corresponds to the bandwidth used for the HFC system at 50% penetration rate in Figure 10.7. The bandwidth of 1.24 Gbps corresponds to the bandwidth that is provided by a SDV system over FTTC for a 500-home node.

POLICY IMPLICATIONS

The main question that policymakers and industry analysts have been asking is whether cable companies have a chance to compete with the RBOCs during and after the convergence of their respective media. The results of this analysis (see Figure 10.7) indicate that cable companies are well positioned to compete with telephone companies. Indeed, Figure 10.7 shows that, in cases where there is small to moderate penetration for interactive services (which would likely be the case in the near term), the costs to the cable companies compare extremely favorably to what the RBOCs would have to spend. To provide a full range of IVSs, cable companies need only to upgrade their existing HFC broadcast network at modest cost (Figure 10.8), whereas a telephone company must install either an FTTC system or an HFC system from scratch.

The FCCs video–dial-tone rules, which may be altered should proposed legislation pass, were intended to oblige telephone companies to build video–dial-tone

(VDT) systems, as opposed to being allowed to build vanilla cable systems. This policy has been undermined, however, by appellate court decisions that have struck down the prohibitions on telephone companies providing cable service. Thus, telephone companies appear to have a choice between offering cable service under Title VI of the Communications Act or VDT service under Title II.

One might ask what the impact of the two regimes is on the telephone companies and, in the long term, on the consumer. The incremental cost of providing VDT service versus plain cable service follows directly from this analysis: It is the difference in cost of an HFC or FTTC system providing switched video services and that of a baseline HFC network providing only analog broadcast video. For an HFC system, adding VDT capabilities can more than double the cost. The FCC justifies the requirement by pointing out that were the telephone companies to build systems capable of only carrying analog broadcast channels, they would not be able to meet the common carrier obligation of serving all comers. However, this requirement means that the RBOCs are being forced to spend a minimum of an additional $100–$300 per home passed to build their networks. In addition, the rule is premised on the assumption that consumers would not mind paying more for the added value. In the long run, however, if consumers cannot rationalize paying more for these advanced services, the telephone companies would be at a distinct disadvantage because the cable companies would be able to offer cheaper rates for the plain vanilla services. However, if the FCC mandate leads telephone companies to install all FTTC systems, they will be better positioned in the event that IVS is highly popular than if they had built a standard HFC system to compete with the cable companies.

If the recent appellate court decisions are not overturned by the Supreme Court, telephone companies may be permitted to offer cable service under Title VI. Should they do so, however, they then become subject to the FCC rules, which require that basic tier cable service be receivable on cable-ready TV sets. This latter requirement effectively forces the telephone companies to build HFC overlay systems if they want to be regulated as cable service providers, even if they build FTTC systems for later provision of SDV.[6] The results of the models indicate that the cost of this requirement is significant, and amounts to as much as $300 per home passed.[7] Without this requirement, the case for an integrated HFC versus an integrated FTTC system would be much less clear. Although an all HFC system still looks most attractive, the price difference is small enough, particularly at high IVS penetrations, that modest improvements in the underlying economics of FTTC, or other considerations, such as long-term maintenance costs, could swing an RBOC to choose FTTC over HFC as the architecture of choice.

---

6. No LEC has yet petitioned the FCC for a waiver under CFR 76.630.

7. The cost of this requirement is the cost of a vanilla HFC plant designed to provide only analog broadcast video. Note also that the typical scenario referred to is a 1,000-home suburban plant, of which 70% is aerial.

The analysis also makes clear that FTTC systems cannot be cost justified to handle voice only. If the RBOCs build voice-only FTTC systems and the regulators allow all of the costs to be put in their telephone service rate base, consumers will be paying a significant premium in initial capital costs to position the LEC to more easily upgrade to video later. This effectively amounts to cross-subsidization of video service provision with telephony service, and is unfair both to consumers and cable companies. Ideally, the incremental cost of adding advanced interactive services later should be paid by those consumers requesting those services and not by telephone rate payers. The current model indicates that the cost to consumers amounts to over $300.[8] If FTTC costs continue to drop, however, the argument could begin to shift, especially because FTTC provides much greater flexibility to handle unexpected future growth than does DLC. With FTTC, the fiber in the distribution plant can easily carry additional traffic, without the need to pull more cable except, perhaps, in the drop plant. With DLC, unexpected growth can require pulling a new copper distribution plant.

This analysis also sheds light on the ability of an HFC system to meet the demand for Internet access. HFC systems cannot readily offer significant upstream bandwidth. For example, if a significant number of consumers want upstream bandwidth for outgoing video from their home-based Web servers, HFC would not be well suited to handle their demands. The same goes for businesses that might want to be content providers and not just content consumers. Thus, from a public policy perspective, if the long-term goal is to empower everyone to be a producer as well as a consumer of content, policymakers should be opposed to HFC systems due to the asymmetric nature of the architecture.

The results of this analysis confirm that there are indeed economies of scope in the provision of interactive video and other broadband services. It is easy to see from Figure 10.7 that it costs significantly more to provide services via overlaid architecture pairs (e.g., HFC over FTTC) than it does via architectures that are upgraded by adding more equipment like electronics.

SUMMARY AND CONCLUSIONS

The results of this analysis clearly indicate that the all HFC architecture is the most economical option for the provision of IVSs, particularly at small penetration rates. However, because the all FTTC option is relatively insensitive to penetration, one can expect it to rival the all HFC system with increasing demand for IVSs. As time goes on, FTTC component price reductions should also narrow the gap between the FTTC-based systems and the all HFC architecture. Moreover, the uncertainty in the all HFC scenario is probably larger on the upside: Managing ingress noise could turn out to be costly over time. The analysis also shows that

---

8. This is the difference in the baseline cost of infrastructure for FTTC versus DLC.

the cost of the all HFC system is highly sensitive to bandwidth requirements—a fact that should make the FTTC-based systems more competitive in the long run. The economics of the five scenarios considered also shows that cable companies have a good chance to compete in the new arena of broadband interactive services. Indeed, the results show that, at low penetration rates, cable companies would be able to offer services at much lower costs than the telephone companies.

This analysis also sheds light on what the FCCs VDT mandate costs the LECs and the rate payers. The results show that this requirement forces the LECs to spend an additional $100–$300 per home passed to build their networks. In addition, this mandate assumes that consumers would not mind paying more money to purchase the advanced services that the LECs would then be able to offer. If this assumption ends up being invalid, the telephone companies would be at a distinct disadvantage. Another important conclusion to be drawn is that the Cable Acts insistence on supporting cable-ready TV receivers currently does not represent a binding constraint. This is because the all HFC system is currently the most economical system. However, the analysis shows that long-term demand increases and price reductions in FTTC components could result in FTTC-based systems that could compete favorably with HFC-based architectures. Under these assumptions, the cable-ready rules, absent a waiver from the FCC, could effectively force an LEC to give up the option of Title VI regulation to be free to choose to build a pure FTTC system.

The results of these models also show that a baseline FTTC for voice has higher initial capital costs than DLC. This means that, should the telephone companies build voice-only FTTC systems and put all the costs into their telephone service rate base, consumers could end up paying more for telephone service—the difference being the added cost for positioning to provide video infrastructure. This would amount to cross-subsidization of video service provision with telephony service provision, and would also be unfair to the cable companies. The counter-argument that is made by the LECs is that FTTC will have lower operating costs due to its enhanced flexibility to respond to requests for second and third lines per home. Last, the results of the analysis indicate that there are indeed economies of scope in interactive service provision, and that, despite their low cost, HFC systems are not well suited to provide interactive services where consumers also produce content. Therefore, in choosing the best technology at the cheapest cost, the FCC should bear in mind that the asymmetric nature of HFC might, in the long term, render it unsuitable for such interactive services.

REFERENCES

Mason, R., Himayat, N., Eldering, C., Omoigui, N., & Sirbu, M. (1994, June). *Overview of Hybrid Fiber-Coax and Fiber-in-the-Loop Architectures.* Presented at the National Fiber Optic Engineers Conference.

Omoigui, N., Sirbu, M., Eldering, C., & Himayat, N., (in press). The economics of competing integrated broadband architectures.

IV

INTELLECTUAL PROPERTY

Service Architecture and Content Provision: The Network Provider as Editor

Jeffrey MacKie-Mason
University of Michigan

Scott Shenker
Xerox PARC

Hal R. Varian
University of California, Berkeley

The formation of the National Information Infrastructure (NII) is often depicted as a battle between various economic entities, such as cable-TV companies and Regional Bell Operating Companies (RBOC's). More recently, the Internet, with its enigmatic origins, anarchic organization, and startling growth, has gained widespread visibility, and is now seen as a fundamental component of the NII. In addition, the various online services (America Online [AOL], Prodigy, CompuServe, etc.), direct broadcast TV, wireless telephony, and other emerging technologies will also likely be important parts of the NII.

These various delivery systems display a wide range of characteristics. The physical transmission medium runs the gamut from satellite transmission to coax cables to fiber optics. Some components are broadcast in nature (everyone in the system receives every signal) and others are switched (signals are directed to specific receivers). The corporate entities deploying these component technologies are diverse from startups to established concerns, and from local monopolies to international competitors and they face disparate regulatory restraints. These physical, corporate, and regulatory issues have been thoroughly discussed in the literature, and thus are not addressed here. Instead, we focus on another issue that, to date, has received considerably less attention.

The purpose of these systems is to transport information content. The interest here lies in how the differences between systems affect the offering of content to consumers. We begin with the observation that these systems have radically different *service architectures*. By architecture we do not refer to the actual physical implementation of the network. Rather, we refer to the nature of the transport service offered. We concentrate on one crucial feature of service architecture: transparency. The entity that transports the bits to consumers is called the network provider. In some architectures, the network provider is aware of the content of transported bits (e.g., movie, point-to-point video, music, etc.); in others, the content is completely opaque to the network provider (see Shenker, Clark, & Zhang, 1994, for a brief and speculative discussion of the implications of this distinction).[1]

We then ask the question: How does this difference in architecture affect the content provided to consumers? Does the difference in the network provider's awareness affect the selection of existing content that is made available to consumers? These are questions that are getting increasing attention in the press concerned with the strategic developments in competition among the Internet, telephony, and cable networks. For example, Gilder (1995) wrote that:

> Networks promote choice, choice enhances quality and quality favors morality. Television is culturally erosive because its small range of offerings requires a broad, lowest-common-denominator appeal. Linking to millions of cultural sources, global networks provide a cornucopia of choices, like a Library of Congress at your fingertips. On the Net, as at a giant bookstore, you always get your first choice rather than a lowest-common-denominator choice. A culture of first choices creates a bias toward excellence and virtue (p. 132).

Also important is whether architecture choice affects the incentives to create new content. This question is addressed in another article (MacKie-Mason, Shenker, & Varian, 1995).

In this chapter we explore one way in which architecture may affect content provision: through the extent to which the network provider can play an editorial role in selecting the content made available to consumers. In an aware architecture, the provider can offer an editorial service; in a blind architecture, it cannot. The different architectures are characterized in the next section. In a subsequent section, we consider the effects of architecture on the selection of already created content to be offered on the network.

---

1. in a later section we clarify that there are gradations of awareness, from awareness of the applications (e.g., teleconference vs. file transfer) to awareness of the actual content (e.g., the specific movie being transmitted).

There are other important ways in which service architecture can affect content provision. A related article (MacKie-Mason et al., 1995) examined two: through technological and institutional delivery costs that vary across architectures, and through the extent to which architecture permits the network to differentiate transport prices for different goods. We discuss these, and some effects on the incentives to create, new content at the end.

The relationship between architecture and content provision is too complex to yield a simple and definitive answer. The purpose here is to provide some initial intuition about the effect of architecture on content provision. To that end, this question is analyzed in the context of some simple models. These are used to identify some of the major issues, and are illustrated through examples. In particular, throughout, this discussion is restricted to the case of a single network provider that serves customers who choose to connect. The present analysis does not address the more complicated, and realistic, case of multiple coexisting and competing architectures.

ARCHITECTURES

The architectural distinctions we make concern the extent to which the network provider distinguishes between the bits it conveys. The Internet, telephony, and cable-TV occupy very different places along this spectrum; they are representatives of the three basic architectural choices we describe.

Architectural Choices

APPLICATIONS-BLIND NETWORKS. One of the Internet's central design principles is that the network provides only bit transportation; it is up to the end hosts to construct higher level applications on top of this raw transport mechanism. This architecture has the feature that it need not be modified as new applications arise; applications are implemented entirely at the end hosts, and no centralized authority need approve such applications.

We use the term *application blind* to refer to architectures where a general interface is made available to end users, who then implement their applications on top of this interface. For purposes of this analysis, we assume that application-blind networks operate as common carriers. They offer a single, nondiscriminatory price for transport, and accept any and all traffic at that price.

Network blindness has both advantages and disadvantages. The ability for end users to develop and implement new applications without network intervention may encourage creativity and experimentation. For example, the Internet has seen a proliferation of applications, such as electronic mail (e-mail) and the World Wide Web (WWW), which were not envisioned when the Internet Protocol (IP) protocol was originally designed. However, a blind network provider is unable to provide a gateway or editorial service that selects which applications and goods to make available to users. For example, the explosion of content on the Internet has led to an increasing amount of low-quality and irrelevant material that users must search through to find valuable content. A network cum editor can filter and certify content, much as the editor of a newspaper does.[2]

APPLICATION-AWARE NETWORKS. Telecommunication infrastructures developed by private enterprise, such as cable-TV and telephony, are more tightly coupled to specific applications. The underlying transport function deep within these networks may be general, but the interface presented to users is highly restricted. In this type of network, the architecture is aware of the type of application being used.

For example, telephone networks are designed around voice traffic, and cable-TV protocols are designed for video traffic. Similarly, online service providers such as CompuServe and AOL generally know what kinds of applications (voice, data, images, etc.) are being transmitted.

An *application-aware* architecture means that the network service provider can identify the general type of application being invoked (e.g., e-mail, audio playback, interactive video). This permits some degree of editorial service. For example, AOL permits users to access *Time* magazine, but for its first several years did not provide access to content on the Internet.

CONTENT-AWARE NETWORKS. Besides being able to determine the general applications customers use, some networks can also monitor and even control the content that is transported over some applications. For example, cable-TV can distinguish basic from premium channels, and Video-on-Demand (VOD) systems know what movie has been requested. Online services also sometimes know what sort of content is being requested; airline prices, bibliographic reference, cartoons, and so on. We define a *content-aware* architecture as one in which the network provider can identify the network content.

It is not the case that all application-aware networks are also content-aware. For example, telephony is content-blind, as is e-mail or WWW browsing.[3] When a network is content-aware, the provider can play a more discriminating editorial role.

---

2. One might argue that indexing, rating, information filtering, and editorial services are emerging on the Internet. But these services merely shift the problem to another level: now the network user must sift through a growing number of filter services to select a preferred one. It is implicitly assumed that the network provider has some advantage in providing an editorial service if the architecture permits it. This advantage might follow from a more credible ability to signal quality, perhaps due to the large fixed investment the network makes in infrastructure.

3. The line between applications and content-awareness is not always clearly defined. For example, an online service might know that a user is accessing *Time* magazine, but may not be able to track which articles the user is viewing. Is this application- or content-awareness? In part due to this blurriness, the two types of awareness are not distinguished in the present analysis.

Architectural Implications for Content Provision

Our fundamental point is that network architecture can have important implications for the nature of information goods made available. Anecdotal evidence is certainly consistent with this view. For instance, the application-blind Internet supports a diverse and rapidly growing set of applications, such as e-mail, file transfer, teleconferencing, and WWW. The application-aware telephone system supports only a narrow range of service (basically FAX, low-speed modems, and telephony), but providers access to varied content in the form of 900 numbers. Cable-TV, which is content-aware, offers only one application and rather limited content. Are these differences related to architectural differences? Because the network provider is the entity actually delivering the goods to consumers, the network providers' policies will affect which goods actually are available to consumers; we investigate the role of architecture in shaping those access decisions.

To focus on the role of architecture, we simplify or eliminate most other relevant factors. For instance, we assume the network providers is a monopoly. This provider is free to maximize profits without competitive (or regulatory) pressures. We also assume that content provision is competitive, with a large number of content providers, and incurs no marginal cost (one can easily incorporate a finite marginal cost into the formalism at the expense of notational simplicity).[4] Therefore, the monopoly network provider can set prices without negotiating with the competitive upstream content providers.

Our modeling of goods and consumers is also quite simple. In most of what follows, separate goods are labeled by an index i, although in some examples it is more convenient to consider a continuum of goods labeled by $x \in [0,1]$. These goods all have equivalent bandwidth requirements. The consumers are modeled as consuming, at most, one unit of each good, with a reservation value v_i^α for consumer α. Initially, we assume that marginal willingness to pay is independent of the consumption of other goods, so a consumer's satisfaction is merely the sum of the consumed v_i^α less the price paid for the goods, and any other nonmarginal costs imposed by the network.

In a content-blind, but application-aware, architecture, different goods denote different applications. In a content-aware architecture, different goods can refer to different content as well as different applications. Given this different definition of *good* in the content- and application-aware architectures, our analysis need only distinguish between aware and blind architectures. The blind architecture cannot distinguish between goods, and the aware architecture can. This key architectural

---

4. With competitive content provision and no marginal content cost, there is nothing to be gained from vertical integration between the network and content providers. Nor could bundling be used to raise profits. We intentionally sidestep these interesting and important strategy questions in this chapter.

distinction has several different implications for content selection from already created goods, which we discuss next.

CONTENT PROVISION

How do different service architectures affect the provision of already created information content? The focus in this chapter is the opportunity for an aware network to provide an editorial service by controlling the content available. We explore two possible motivations for limiting content: *clutter* costs and *attention* costs. Clutter cost is an increase in the difficulty of finding or processing information that results from the total number of information goods available on the network. Such costs arise at the network level; they are not attributable directly to individual consumer decisions. Attention costs are a reduction in the value of information goods as the number of goods purchased by a users increases. These costs can be traced to decisions made by individuals.

Costs of clutter and attention begin to distinguish the economics of information goods from other more traditional fields in economics. Of course, there is nothing novel about externality costs or interdependent demands, which are the formal characterizations used to specify clutter and attention. However, in the context of information networks, much of this chapter's attention is directed to whether there are too few or too many goods offered. This is fairly unconventional for an economic problem; more choice over available goods is routinely assumed to be unambiguously desirable. Questions of this sort do appear in the economics of advertising, but research in that area has also been somewhat unconventional, and plausibly is an early example of the economics of information.[5]

The effect of different types of cost on content provision depends on the network architecture. An aware network with its greater control is advantaged when there are significant network costs (i.e., the selection of which goods to provide is more efficient). There is no such advantage with user costs because these can be allocated efficiently by the actions of individual users without network intervention.

The different locus of control also implies a difference in the selection order for content. An aware network controls content selection, and thus orders choices by the profits each good generates. In a blind network, with its single transport price p, customers control content selection. Any good with positive demand at price p will be purchased by some consumer, so goods are generally ordered by maximal willingness to pay.

---

5. There is quite a bit of research on the optimal variety of goods when products are differentiated, but that literature is concerned with closely related substitutes. We investigate markets with different information goods. Indeed, when clutter costs are modeled, we assume that demands for different goods are strictly independent.

This contrast between ordering by profitability versus maximal willingness to pay has interesting implications for the content diversity on various networks. For example, consider two different kinds of goods: low-value *mass-market* goods, which have low maximal willingness to pay but high total revenue, and high-value *niche* goods, which generate relatively little revenue but have high maximal willingness to pay. Aware architectures will tend to favor low-value mass-market goods, whereas blind architectures will favor high-value niche goods (see MacKie-Mason et al., 1995, for further discussion). The bias toward mass-market goods is consistent with what is observed when comparing the offerings of the Internet with cable-TV.[6]

We make these observations more concrete by considering how clutter and attention costs affect the provision of goods.

Clutter Effects

When there are many information goods or applications available, the clutter that results can decrease the value to consumers. It becomes harder (and slower) to locate the desired content, or the interface becomes more difficult to use.

Blind networks cannot directly control the variety or quality of content available. They only set a single, uniform transport price that acts as a cutoff. Any good or application worth more to some user than the uniform cutoff price will be offered, regardless of the clutter costs imposed on other users. The concern with the resulting clutter in blind architectures is evidenced by the huge popularity of Web indexing services (http://www.yahoo.com/ receives about 6 million visits per day) and the demand for them from commercial providers (e.g., AOL recently purchased WebCrawler).[7] Readers of Usenet newsgroups and Internet mailing lists are also familiar with clutter costs. The current Usenet feed is about 5MB of new, mostly unmoderated text material per day.

Content-aware architectures can control the content available to users and play an editorial role. In fact, this is already standard practice in the moderated discus-

---

6. This is not to imply a causal link; there are many other differences between these two systems that could explain this observation. For instance, the broadcast nature of cable-TV would favor mass-market goods while the switched nature of the Internet does not, although the advent of multicast is changing that somewhat.

7. The field of Web indexing services is getting quite cluttered: There are at least a dozen competing services available. To help users choose an index are at lest a dozen competing services available. To help users choose an index service, so that they may then use the index to help them choose content, several comparative studies of index services have been conducted (see e.g.,http://www.zdnet.com/pc-comp/features/internet/search/index.html).

There are several meta-search services (that search multiple search services), such as the All-in-One Search page, CUSI, Internet Exploration Page, and SavvySearch. That is, there are already editorial meta-services that recommend editorial services.

sion groups of many online services. Online services also restrict the variety and quantity of content available.[8]

Clutter costs are network costs: They depend on the total number of applications or goods available, not the choices made by particular individuals. However, these costs arise not as expenses directly incurred by the network in providing a service, but in reduced user utility as more goods are offered. We model clutter effects by decreasing an individual's utility as more goods are offered (see Shiman, 1995, for a discussion of similar congestion effects for e-mail). For instance, if the network offers a set of goods G and a consumer α has purchased a set G_α of goods, the total valuation by that consumer is $\Sigma_{i \in G_\alpha} v_i^\alpha - F_c(|G|)$, where F_c is some nondecreasing function. If $\Sigma_{i \in G_\alpha} v_i^\alpha - F_c(|G|) < 0$, one assumes the consumer leaves the network. Let C denote the set of connected consumers, those for whom $0 < (\Sigma_{i \in G}(v_i^\alpha - p_i)_+ - F_c(|G|))$.

A network with a blind architecture sets a uniform transport price p. Let the revenue to a content provider for good i be $R_i(p)$. Then provider i will offer its good if and only if $R_i(p) > 0$. However, adding a good increases clutter costs for all network users; some marginal customers may disconnect. Thus, revenue must be computed over those consumers still attached to the network, $\alpha \in C$. From this one can make the following observation about content provision in a blind network with clutter costs: The selection of goods is ordered by maximal willingness to pay (i.e., any good j for which $\max_{\alpha \in C}[v_j^\alpha] > p$ will be offered). Thus, *the users* select the content available, rather than delegating the selection decision to the network.

In the presence of clutter, a network provider with an aware architecture can decide which goods to make available. The problem is to choose this set G and the associated p_i to maximize the total revenue, which is given by the sum $\Sigma_{\alpha \in C} \Sigma_i p_i \delta(v_i^\alpha > p_i)$.[9] This maximization problem depends on the details of the distribution of consumer preferences. The striking difference from the blind network is that the goods are ordered by maximal revenue, rather than by maximal willingness to pay.[10] This is the natural consequence of the locus of editorial control: The aware net makes a profit-maximizing choice for the network of users as a whole; in a blind network, each individual user self-selects the desired content.

---

8. Interestingly, the online services have recently engaged in a mixed strategy, providing both an editorially controlled service and a gateway to unrestricted Internet content. This evolution is discussed later.

9. The delta function is defined $\delta(z) = 1$ if z is true, 0 otherwise.

10. The revenue again must be evaluated with respect to those consumers still connected to the network, given the other goods offered. We formalize the notion of profit ordering by using a necessary condition for offering good j, conditional on the set of other goods that are offered at an optimum. The incremental good j is sold to some subset of connected users, yielding revenue $R_j(p_j)$. However, some users detach from the network as a result of the extra clutter. Let D_j denote the set of users who detach—that is, those for whom $F_c(|G_{-j}|) - F_c(|G|) - v_j^\alpha \delta(v_j^\alpha > p_j) > S_{-j}^\alpha$, where G_{-j} is the set of offered goods excluding good j, and S_{-j}^α is the surplus obtained by user α from goods G_{-j}. Then one can say that goods are ordered by revenue in the sense that all offered goods satisfy $R_j(p_j) > \Sigma_{\alpha \in D_j} \Sigma_{i \in G_\alpha} p_i$, and all other goods do not.

The different ordering of content selection can be interpreted as a bias toward mass-market goods in an aware network. Offering a new good has an opportunity cost: the revenues lost from customers who detach due to the increased clutter, plus the revenue lost from lowering some prices to keep other customers attached. This opportunity cost must be overcome for a new good to be offered. Consider two new goods with identical value to all users, but that appeal to different fractions of the user base. The mass-market good is purchased by f_m users, the niche good by f_n users, and $f_m > f_n$. The revenue from the mass-market good with f_m customers will be greater than the niche good revenue, and it will more likely exceed the opportunity cost. Indeed, even if the niche good were more highly valued by users, the aware network favors the mass-market good if the difference in market size is large enough: $f_m p_m > f_n p_m$.[11]

When there are clutter costs, a network with an aware architecture offers different goods than does a blind network, but does it handle clutter better? Clutter costs are an externality: Offering a new good benefits some users, but hurts all others. In a blind network, content providers will decide to enter the market as long as there is a single user with $v_j^\alpha > p$; the resulting clutter will tend to reduce total consumer surplus. This is the typical tragedy-of-the-commons phenomena present in many congestion models (see MacKie-Mason & Varian, 1995, for an analysis of congestible resources).[12] In general, an aware network provider will not only select different goods to offer, but will also limit the number of goods offered to reduce total clutter costs.

This is easy to see in the case when all users who purchase a good value it the same—then the aware network provider charges a price equal to this value for each good. To induce the consumer to connect, the provider pays a subscription rebate equal to the user's clutter costs, $F_c(|G|)$. In this way, the network provider extracts exactly $\Sigma_{i \in G} \alpha v_i^\alpha - F_c(|G|)$ from each consumer α, or the total consumer surplus. Because the aware provider is maximizing total surplus, it selects the first-best set of goods—the clutter externality is internalized. More generally, the network will not achieve the first best. However, reductions in consumers' surplus, due to clutter, reduce the profit it can extract, so the aware provider acts as an editor by limiting the number of goods offered to reduce total clutter costs.

---

11. Although surely not due to clutter alone, the reports that about one quarter of users of online services like AOL detach within a year suggests that the number of marginally attached customers is sufficiently large that this bias in selecting new content could be important (see MacKie-Mason et al., 1995, for a more complete discussion on the mass-market bias in aware networks).

12. Another aspect, which is not explored here (in fact, it is precluded by the particular modeling of the clutter effect), is that if different goods experience different amounts of clutter (i.e., are devalued differently), then goods that are less susceptible to clutter can displace those that are more susceptible to clutter, even if their intrinsic value is less. This is akin to the effect that, in a computer network, congestion-tolerant traffic can squeeze out congestion-intolerant traffic, even though the congestion-intolerant traffic may intrinsically be much more valuable.

The clutter cost externality has important implications for interpreting the different content orderings discussed earlier. Recall that the aware architecture will favor mass-market over niche goods (appropriately defined). This is not necessarily a bad thing. A *niche good*, by definition, benefits a small subset of users while imposing additional clutter cost on all others. On net, total welfare may be lessened by introducing a niche good. The appendix shows that, indeed, too many niche goods will be available on a blind network.

We have characterized two effects that service architecture has on content provision when there are clutter costs. First, an aware network offers content based on a profit ordering; content is offered on a blind network according to maximal willingness to pay. Second, the clutter cost is an externality that can be at least partially ameliorated by an aware netwrok provider that acts as an editor or gateway. We now illustrate these aspects of the clutter effect through two numerical examples. A more general treatment is presented in the appendix.

Example 1: No Blind Equilibrium.   There is a continuum of goods labeled by x, with $x \in [0,1]$. Each individual consumer α values each of these goods an amount v^α, and these values v^α are uniformly distributed between 0 and 1 throughout the population.

Let p be the transport price of these goods (which are priced the same even in the aware architecture because they are essentially identical). Suppose a network offers x goods; if $v^\alpha > p$, then user α connects and buys all x of the goods, to receive utility

$$U_\alpha = \int_{1-x}^1 (v^\alpha - p)ds - F_c(x)$$

It is assumed that the clutter effect function F_c is of the form: $F_c(z) = \lambda z^2$. By solving for $U_\alpha = 0$, one finds that all users with $v^\alpha > p + \lambda x$ are connected; thus, the fraction of connected users and the total demand for each offered good is $1 - p - \lambda x$. Total profit is calculated as demand times prices times the number of goods, or $\pi = (1 - p - \lambda x)px$.

An aware network can set both p and x to maximize profit; the maximizing values are $p = {}^1/_3$ and $x = {}^1/_{3\lambda}$, yielding a demand of ${}^1/_3$, a revenue of ${}^1/_9$, and a consumer surplus of ${}^1/_6$ per unit good. Goods are offered as long as they yield revenue at least ${}^1/_9$; at $x = {}^1/_{3\lambda}$, any additional goods would yield less revenue. Although the example is stylized, in that each good is essentially the same, it illustrates the result that goods are selected based on a profit condition in an aware network.

A network with a blind architecture can set p, but not x. Content providers will enter the market as long as some customers are willing to buy. For any p, as long as the demand per good $1 - p - \lambda x$ is positive, new goods will enter, ultimately driving the demand to zero. Thus, in the blind architecture, no stable equilibrium can have positive demand, revenue, or consumer surplus. Given the set of offered goods, a new good will be purchased (and profit will be positive, so it will be offered) as long as at least one customer values it more than the transport price—that is, if $\max_\alpha v^\alpha > p$. Because the optimal price is $p = {}^{1-\lambda x}/_2 < 1$ and $\max_\alpha v^\alpha = 1$, all goods are offered until congestion crowds out all users. This illustrates both that goods are selcted according to maximal user value in a blind architecture, and that clutter creates a congestion externality.

Example 2: An Inefficient Blind Equilibrium. Consider an example with a continuum of goods labeled by x, with $x \in [0,1]$. Each individual consumer values each of these goods an amount x (i.e., the x'th good is valued an amount x). The clutter effect function F_c is: $F_c(z) = {}^1/_2(z - {}^1/_2 + \varepsilon)_+$. In the blind architecture, all goods have the same price p, and all goods with $x \geq p$ are offered. Because $R = p(1 - p)$, revenue is maximized at $p = {}^1/_2$, and the resulting revenue is $R = {}^1/_4$. The consumer surplus is $S = {}^1/_2(1 - p)^2 - {}^1/_2({}^1/_2 - p + \varepsilon)_+$, which takes on the value ${}^{4\sqrt\varepsilon}/_8$ when $p = {}^1/_2$. Then, for sufficiently small $\varepsilon > 0$, the total welfare and consumer surplus are increasing in the region ${}^1/_2 < p < {}^1/_2 + \varepsilon$, even though the blind revenue is maximized at $p = {}^1/_2$. This last statement can be verified by noting that the price derivative of the total welfare is $D'(p)(p - F'(D(p)))$, where $D(p)$ is the demand function. Thus, the total welfare increases with p if and only if $F'(D(p)) > p$.

In this example, the clutter externality in a blind network can be resolved with a traditional Pigovian tax to raise the price by ε. As an alternative, an aware network fully internalizes the clutter cost and achieves the first best. Of course, consumers would prefer the blind network because they get to share in the surplus, all of which is extracted by the network provider before the architecture is aware.

Summary

In the appendix we present a more general model of content provision with a clutter externality. In that model, a Pigovian tax may lower welfare in a blind architecture, in contrast to Example 2. However, the editorial service provided by the aware network internalizes the clutter problem. It is formally shown here that a blind network will offer too many goods, and that the first-best involves eliminating some niche goods.

In this analysis of content selection with clutter costs, we have uncovered two principles:

1. *Externality*: Content value is reduced as the menu of offerings becomes more cluttered. An aware network provider can serve a beneficial editorial function, increasing the value of the net by limiting the offerings.

2. *Content Ordering*: Because the aware net provider controls the offerings, content selection will be ordered by profit under an aware architecture, and by willingness to pay under a blind architecture.

Suppose that network providers or others develop ways to control clutter cost even in a blind architecture. Even so, an individual user has limited time and attention to devote to different information goals. The next section explores how network architecture affects content provision when users experience attention costs.

Attention Effects

In the previous subsection we explored the fact that, when a consumer purchases many goods, his or her satisfaction may be significantly less than the sum of his or her v_i^α's. It is assumed that the whole may be less than the sum of the parts due to clutter: The more goods or applications available, the harder it is for the user to find what he or she wants. Another possibility is that users have limited attention, and their enjoyment of a given good is decreased when they are also consuming other goods. There has been considerable recent discussion of the extent to which we are moving from a service economy to an attention economy (see e.g., Lanham, 1993). For example, a subscription to HBO has reduced value if the user also subscribes to additional movie channels. This is a special case of the goods being (imperfect) substitutes for each other. Attention depends on the goods *consumed*, whereas clutter results from the goods *offered*.

It might appear that attention effects are a modest variant on clutter effects, and not worth the bother of separate analysis. Indeed, we initially conjectured that the same would hold: that content selection would be ordered differently in aware and blind networks, and that the aware architecture would have an advantage because it could limit the attention cost imposed on users.

In fact, both of these conjectures turned out to be wrong. With a little thought, it may be obvious why they are wrong, However, it is worthwhile to explore how attention effects differ from clutter effects. For one thing, demand interactions of this sort surely are relevant for an aware net provider selecting content (although not in the way we originally guessed). In addition, the economics of service architecture is a novel topic, and it is instructive to understand how slight differences in modeling assumptions can lead to quite different results.

No Externality.   The first thing to notice is that the attention effect does not create an externality. Attention cost is not imposed on users: They impose it on themselves. Suppose individual utility is $\Sigma_{i \in G\alpha} v_i^\alpha - F_a(|G_\alpha|)$ for a consumer who purchases a set of goods, where $F_a(\cdot)$ is the nondecreasing attention cost of consuming those goods. The $F_a(\cdot)$ term represents interdependence in a single user's utility function; it is unaffected by what goods other users consume. Thus, there is nothing intrinsic about the attention effect that favors the aware architecture over the blind.

We use a simple example with homogenous users and heterogeneous goods to illustrate the absence of an attention externality in a blind network. Consider a continuum of goods labeled $x \in [0,1]$, where all consumers agree that x is the value of the good. In a blind architecture, a user will purchase all goods that have value $x > q \geq p$. The cutoff q will generally be greater than the price p because, for each good consumed, the user pays both p and an incremental attention cost, F'_a. Therefore, surplus is

$$U = \int_q^1 (v - p)dv - F_a(1 - q).$$

The attention cost function is $F_a(z) = \frac{1}{2}(z - \frac{1}{8})_+$.

The consumer chooses to purchase all goods with value greater than $q = \frac{1}{2} + p$, where the $\frac{1}{2}$ is simply the marginal attention cost. Substituting back into the surplus function and taking the deriviative with respect to p we find that

$$\frac{\partial U}{\partial p} = \begin{cases} -\frac{1}{2} + p, & \text{if } p \leq \frac{3}{8} \\ -1 + p, & \text{if } \frac{3}{8} < p \leq 1 \end{cases}.$$

For all feasible values of the transport price, surplus is decreasing in price. There is no externality, and reducing the set of offered goods (by raising price) can never make consumers better off.[13]

Content Selection by Value or Profit?   The attention effect is a function of consumed goods, not offered goods, and users choose the goods consumed. As a result, it turns out that in both architectures the goods are generally ordered by maximal willingness to pay, rather than total revenue. Why is this result different from the ordering by revenue for the aware architecture with clutter effects? The clutter cost is a per good cost borne by the network as a whole; it lowers the total consumer surplus available for extraction. This network-wide cost from offering an incremental good must be recovered. In contrast, attention cost is a per good cost borne by individual users. Because there is no network-wide cost to be recov-

---

13. It would never make sense to eliminate a different set of goods because all consumers agree on the value ordering of the goods. Thus, raising price eliminates those goods that are valued least by all consumers.

ered, the network offers goods based on individual consumer valuations, not total revenue.

As before, the result is trivial for a blind network: The net provider merely sets the uniform transport price p, and all goods that some consumer values more than p are offered (consumer valuation is net of the marginal attention cost induced by consuming the good). For an aware network, the result is also straightforward. The net provider should offer all goods for which $\max_\alpha[v_i^\alpha - F_a'(|G_\alpha|) > 0.$[14] To see why this is so, suppose not for good j: If the network adds j to its offering, at least one user would experience higher utility by purchasing it, and there are no external effects on other users, so no one detaches. Thus, the net could charge a $p_j > 0$ and increase its profits.

It may be worth noting that ordering by willingness to pay is equivalent to ordering by profit in this case. The point is not that the network does not order by profit—of course it must if it is maximizing profits—but rather that with only attention costs the two orderings are identical.

Architecture Choice and Social Welfare

We have characterized the effects that choice of service architecture have on content provision have been characterized. What are the welfare consequences of architecture choice? Given the focus on the selction of goods offered, it is natural to first compare how many goods are made available under different architectures. However, because different goods will be valued differently, we are also interested in how content selection affects the total surplus obtained (i.e., the sum of profits and conumers surplus).

Consider first a network with attention costs. The aware provider can charge a different price for each good or application; the blind provider chooses only one price. As a result, profits will be higher and more goods will be offered in an aware network. The aware network number of goods is greater becasue all goods are offered for which $\max_\alpha[v_i^\alpha - F_a'] > 0$, whereas in the blind net only goods for which $\max_\alpha[v_i^\alpha - F_a'] > p > 0$ are offered.

How do the number of goods provided by aware or blind monopoly network providers compare to the first-best welfare optimum? It is easy to determine the welfare-maximizing outcome when there are only attention costs. Each user orders all of the goods from the highest v_i^α to lowest. The user then adds the goods to his or her consumed set in order until $v_j^\alpha < F_a'(j)$. The union of the goods desired by all users is the optimal offered set. That is, all goods for which $\max_\alpha[v_j^\alpha - F_a'(|G_\alpha|)] > 0$ should be offered. This is the set of goods offered by the aware net-

---

14. The notation is abused slightly by referring to the incremental change in attention costs as the derivative F_a'. In fact, the argument changes by integer values, and the incremental change is a first difference.

work, so the aware provider makes the optimal selection, whereas a blind network offers too few goods.

Although the aware architecture offers the optimal number of goods when there are attention costs, both architecures achieve less than the optimal level of social welfare. For the blind architecture, the result follows directly from offering too few goods. In an aware network, the provider chooses prices $p_i > 0$. Now consumers order the goods by $v_i^\alpha - p_i$. Thus the order in which the consumer selects goods is distorted. Further, the consumer purchases goods only until $v_j^\alpha - p < F'_a$, so an individual consumer takes fewer goods than the social optimum $(v_j^\alpha < F'_a)$. That is, the right set of goods is offered, but too few consumers buy them. This result contrasts strongly with the result for a network with clutter costs. In that case, the provider used its editorial opportunity to offer fewer goods under an aware architecture.

Although network pricing lowers social welfare when there are attention costs, there is no market failure, as with the clutter effect. The result is simply the usual monopoly result: The net provider restricts the output to raise prices above marginal cost and earn supracompetetive profits. The editorial role in an aware network is not an intrinsic advantage when there are attention costs: Private consumers internalize the problem and solve it themselves.

Both architectures offer suboptimal social welfare when there are attention effects. Which one does better? Unfortunately, this cannot be determined. Although more different goods are offered in an aware net, $p_i < p$ for some i, $p_j > p$ for some j, and the net effect on social welfare is ambiguous.[15]

What happens to the number of goods offered and social surplus when there are clutter costs rather than attention costs? It was shown previously that the blind architecture offers fewer goods than an aware network. This provides another sharp contrast between clutter and attention effects because a blind network offers more goods in the latter case.

How does an aware network compare to the social optimum when there are clutter costs? It was shown earlier that when all purchasers of good i have the same valuation for that good, $v_i^\alpha = v_i^\beta$, the aware provider could achieve the first best by charging $p_i = v_i$ for the goods, and then giving a subscription rebate to users to cover their clutter costs (assuming that everyone experiences the same clutter costs). Now suppose there is some variation across users in the clutter cost function. Then some users will get positive surplus from the original subscription rebate, whereas others will want to detach. To retain some of the surplus from those users who want to detach, the network will exclude some low-profit goods to reduce clutter costs. Thus, one would expect to see too few goods on an aware network with clutter costs, consistent with the usual monopoly result. It is not certain whether the blind net offers too few or too many goods relative to the optimum

---

15. That is, more different goods are offered, but less is consumed of some of the goods.

because the monopoly profit incentive and the clutter externality effect work in opposite directions.

Unfortunately, the effect of architecure choice on total surplus is ambiguous when there are clutter costs, as it is for attention costs. In general, for both architectures, welfare will be lower than the optimum if the network provider is a monopolist. However, either the blind or the aware architecture could surpass the other, depending on consumer preferences.

PRICE DIFFERENTIATION AND PROVIDER COSTS

Another article (MacKie-Mason et al., 1995) explored other ways in which the difference in service architecture can affect both content provision and the incentives to create new content. We summarize some of the main results here to flesh out a richer view of the economics of service architecture.

Suppose there are no clutter costs, attention costs, or other consumer disutilities. Then the main feature of the difference in architecture is the fact that an aware network can charge a different price for transporting different goods or applications, whereas a blind network is limited to a single price. One element of provider cost is ignored in the present chapter: There is a fixed cost per good offered that a network with an aware architecture must pay. This is referred to as a *gateway* or *liability cost*. For example, the aware architecture might have to reprogram to add a gateway to deliver a new application. Alternatively, under current and emerging U.S. law (at least), an aware network provider may be liable for certain types or content (e.g., libel or obscenity) that it transports.

When we compare the aware and blind architectures under these conditions, we find that potentially severe inefficiencies and political economy conflicts can arise. For example, if gateway costs are relatively low, the monopolist net provider will generally prefer an aware architecture (because it can imitate blind price as one option, or do something better). But consumers will often (but not always) prefer the blind architecture because they can retain more surplus. Total welfare could be better under either architecture, depending on the specifics.

For example, suppose most consumer variation in v_i^α is in i (i.e., "within" variation, with most users having the same preferences, but each having widely varying valuations across goods). Then the aware architecture has higher total welfare (and network profit), but consumers prefer the blind architecture. However, most variation is "between" different valuations for the same good by different consumers then both consumer surplus and total welfare can be higher under a blind architecture.

When we examine the creation of content, we find an interesting problem of commitment. The aware network can extract all of the surplus from a new information good, whereas the blind architecture leaves more surplus as a reward and thus incentive for the creator. Therefore, one might expect the blind architecture to favor creation. However, the aware network can mimic the pricing policies of the blind network. Therefore, it might seem that an aware network should be able to induce at least as much creative effort as blind network. However, the mimicry strategy is not compelling here because the problem is dynamic: The aware network cannot credibly commit that it will not expropriate the surplus in a future period, after the investment in creation activity is sunk. By choosing a blind architecture, then, a network may be making a credible commitment to leave incremental surplus in the hands of creators, and thus it may induce a higher steady-state level of creative activity.

There is another obvious difference that is likely to affect content creation: In an aware network, a single firm (the network provider) effectively decides whether to invest in creating new content, whereas multiple firms each make independent decisions in a blind network. We expect a blind network to have more experimentation with content creation, and collectively more risk-taking (not because of differences in risk aversion, but because of differences in beliefs about the likely success of projects).

DISCUSSION

One key distinction between various competing visions for the NII is the extent to which the network provider is aware of the content of the bits it is conveying to consumers. Aware architectures are aware of the content, blind architectures are not. The focus in this chapter is the impact this architectural distinction has on the provision of content.

The most striking difference between networks with aware and blind architectures is that the selection of offered goods proceeds by maximal profit in aware networks, whereas goods are ordered by the maximal willingness to pay among users in a blind network. In an aware network this ordering favors mass-market over niche goods. These results apply, formally, when there are either clutter or attention costs. However, when the costs take the attention form (and if there are no other network-wide costs from offering an incremental good), the profit and willingness-to-pay orderings coincide.

Although the selection orderings coincide when there are only attention costs, the number of goods offered under the different architectures do not: An aware network will offer more goods when there are attention costs. This occurs because the aware network provider can differentiate prices between goods, and thus will offer some low-value goods that do not pass the uniform price threshold on the blind network.

In striking contrast, when the costs are experienced as clutter, a network provider with an aware architecture offers fewer goods than a blind network. This is the consequence of our third major finding on content provision: Clutter costs are an externality, but they can be internalized if the architecture is aware. Therefore, in a blind network with clutter, too many goods are offered; the aware network provider exercises its editorial capability to reduce the number of goods offered.

We have presented a simplified and stylized model of network service architecture and content provision. We plan to explore the effects of service architecture in richer settings. For example, one would expect more than one network provider to compete. Then it is easy to imagine that some customers who find clutter very costly will choose to subscribe to an aware network that controls clutter, at the cost of receiving mostly mass-market goods. Other networks might cater to customers with niche tastes, but with sufficiently high values for these niche goods to overcome the resulting clutter costs.

However, once one begins to think about multiple networks with multiple architectures and different menus of available content, one has to take seriously another feature of information economies: positive network externalities. That is, the value of belonging to a network for many consumers tends to increase the more other users who are connected. For example, e-mail is much more valuable if it can be exchanged with anyone, not just with subscribers on one of several competing networks. Therefore, there will be a growing demand for interoperability. Interoperability between multiple proprietary networks with multiple incompatible architectures is costly and difficult: Interoperability seems to favor the blind architecture. Thus, rather than multiple competing network "clubs," there might eventually be a unified internetwork with a blind architecture, on which customer types who suffer high clutter costs pay for editorial services provided by competing sellers over the blind network. Such editorial services may not be as cost-effective or able to build reputation as editorial control provided by a network provider, but they may be the second-best compromise that results from the value of interoperability. Our casual observation suggests this vision is consistent with the recent movement by proprietary online services toward providing a gateway to blind Internet services, while attempting to differentiate their products through the quality of their competing editorial services.

ACKNOWLEDGMENT

This work was supported by National Science Foundation grant SES-93-20481.

APPENDIX: CLUTTER AS AN EXTERNALITY

One key aspect of the clutter effect is that the offering of a good affects all consumers negatively (actually, it only affects those consumers who remain connected), but may offer only some users benefit. This negative effect can be controlled by the aware architecture, but not in the blind architecture.

Suppose there are many possible goods, and that each good appeals to an entirely separate group of like customers. That is, for each good i, there are f_i potential customers, each of whom value the good at v_i, and no consumer wants more than one good. Order the index numbers for the goods so that $v_i > v_{i+1}$. Everyone bears the same clutter costs, which depend solely on the number of goods offered: $F(N)$.

In a blind architecture, the network provider can set single transport price. Customers will participate in the network only if their surplus is positive, so an individual rationality (IR) constraint can be obtained for each group of the form $v_i - F(N) - p > 0$, given N. Given the ordering of the goods, $N = \arg \min_i[v_i - F(N)]$, so the IR constraint can be binding for the least valued good, at most. A profit-maximizing network will thus set $p = v_N - F(N)$, conditional on its choice of N. That is, the price will be set to extract all of the surplus of the group with the lowest surplus. The network then chooses N to maximize its profits where the profit function is $\pi(N) = [v_N - F(N)] \sum_{i=1}^{N} f_i$.

Now some welfare consequences of the clutter externality can be considered. In general, one would expect that, because clutter imposes an externality, there will be too many goods in equilibrium. The first way to investigate this conjecture is to ask whether welfare increases if the price increases. Consumer surplus at price $p(N - j)$ is

$$
\begin{aligned}
S(N - j) &= \sum_{i=1}^{N-j} f_i[v_i - F(N - j) - p(N - j)] \\
&= \sum_{i=1}^{N-j-1} f_i[v_i - v_{N-j}]
\end{aligned}
$$

where the simplification results from substituting in $p(N - j) = v_{N-j} - F(N - j)$. The change in consumers' surplus from raising price from the profit-maximizing level $p(N)$ to any higher price that is profit maximizing for the smaller number of offered goods is:

$$
\begin{aligned}
S(N - j) - S(N) &= \sum_{i=1}^{N-j-1} f_i[v_i - v_{N-j}] - \sum_{i=1}^{N-1} f_i[v_i - v_N] \\
&= -\sum_{i=1}^{N-j-1} f_i[v_{N-j} - v_N] - \sum_{i=N-j}^{N-1} f_i[v_i - v_N] < 0.
\end{aligned}
$$

Thus, forcing the network provider to charge a higher price (say, by imposing a tax) and discouraging some low-value goods from being offered will, in fact, lower consumer surplus, despite the clutter externality. It is already known that a higher price will also lower profits; thus it follows immediately that a higher price will also lower total welfare.

This does not mean that the optimal set of goods is being offered, however. Suppose it were possible to pick and choose which goods would be offered, which is precisely what an aware network can do. How would total welfare be changed by eliminating a single good? Let G_{-j} refer to the set of goods when the jth good is removed from the profit-maximizing set. The change in total welfare from removing the good is

$$
\begin{aligned}
W(|G_{-j}|) - W(N) &= \sum_{i \in G_{-j}} f_i[v_i - F(N-1)] - \sum_{i=1}^{N} f_i[v_i - F(N)] \\
&= -f_j[v_j - F(N)] + [F(N) - F(N-1)] \sum_{i \in G_{-j}} f_i
\end{aligned}
$$

The first term reflects the loss of surplus from excluding the jth group of customers; the summation reflects the reduction in clutter costs that offering the jth good imposes on all other customers. When the clutter savings are large relative to the surplus from the jth good, welfare would increase by excluding the good. This will tend to be the case for niche goods (small f_j small relative to $\Sigma_{i \in G} f_i$), and, of course, when the marginal clutter cost $F(N) - F(N-1)$ is large.

Does an aware network necessarily deal better with the congestion problem? Yes. Aware profits are maximized by extracting the full consumer's surplus with $p_j = v_j - F(N)$, so the first term in the expression is the negative revnue from good j, and the aware network drops goods with the lowest revenue until this foregone revenue exceeds the additional revenue (surplus) that can be extracted from other users as clutter decreases. Thus, the aware network completely solves the externality problem, and the welfare-maximizing set of goods is offered.

As we have shown, when there are clutter effects, an aware network will order content selection by maximal profit, whereas a blind network will order by maximal willingness to pay, just as with the liabilty/gateway effect. However, with a clutter externality, too many goods will be offered in a blind architecture, particularly too many niche goods. Forcing the network provider to raise its transport price may not solve the clutter problem, and, in fact, may reduce both consumer surplus and total welfare. Adopting an aware architecture will internalize the clutter externality, and, in the special case in which the network can extract all surplus, will even result in the socially optimal set of goods.

REFERENCES

Gilder, G. (1995, December). Angst and awe on the Internet. *Forbes ASAP, 4*, pp. 112–132.

Lanham, R. A. (1993). *The electronic word: Democracy, technology, and the arts.* Chicago: University of Chicago Press.

MacKie-Mason, J. K., Shenker, S., & Varian, H. R. (1995). *Network architecture and content provision: An economic analysis.* Technical report, University of Michigan.

MacKie-Mason, J. K., & Varian, H. R. (1995). Pricing congestible network resources. *IEEE Journal of Selected Areas in Communications, 13*(7).

Shenker, S., Clark, D. D., & Zhang, L. (1994). Services or infrastructure: Why we need a network service model. In *Proceedings of Workshop on Community Networking*, pp. 145–149.

Shiman, D. (1995). *When e-mail become junk mail: The welfare implications of the advancement of communications technology.* Technical report, State University of New York-Oswego.

12

Corporate Free Speech Rights and Diversity of Content Control: An Emerging Dilemma

Joseph M. Foley
The Ohio State University

Recent cases have established a fundamental First Amendment right for video system operators. The Supreme Court has suggested that the cable operators' First Amendment rights must be given substantial weight in determining the constitutionality of mandatory signal carriage rules. Lower courts consistently have held that a total ban on the provision of video services violates the constitutional rights of telephone companies. This renewed emphasis on the rights of system operators as the litmus test for policy will have a profound impact on the development of these new media. Policy options for influencing the content in the emerging multichannel media will be constrained by these rights. The policy environment of broadcasting, with its emphasis on licensing and public service, is very different from the approach that will result from a focus on the rights of system operators. With the continuing wave of mergers and acquisitions in the industry, it appears likely that the entire control of content on the media sources most available to the public could rest in the hands of a few corporations. The dilemma and the challenge is to find ways to provide for diversity of content, while recognizing the free speech rights that are being emphasized for those corporations. This chapter seeks to identify policy options that could be pursued in light of the initial court opinions, which have begun to define those rights. By focusing on issues of free speech, rather than economics and antitrust, communication corporations have found a way to make fundamental changes in the way in which they are regulated. This creates a new legal environment for the development of regulatory policy.

CHANGING TECHNOLOGY AND CHANGING CONSTITUTIONAL MODELS

The traditional First Amendment models of common carriers and broadcasting are being replaced by a new model whose characteristics are just beginning to emerge. Although past policies have been shaped by reliance on those models, the rapid economic and technological convergence of the media has undercut the distinctions. This has been accompanied by a more fundamental social change, in which individual rights have come to have greater emphasis than collective rights. Post (1995) pointed to this fundamental shift in constitutional law:

> For the past sixty years American constitutional law has regarded democratic community as a complex dialectic between two distinct and antagonistic, but reciprocally interdependent forms of social organization, which I call "responsive democracy" and "community." I define these forms of social organization in terms of the hermeneutic project of the law. When the law attempts to organize social life based on the principle that persons are socially embedded and dependent, it instantiates the social form of community. When it attempts to organize social life based on the contrary principle that persons are autonomous and independent, it instantiates the social form of responsive democracy... Although the principles of responsive democracy and community often conflict in the outcomes they require for specific cases, American constitutional law has nevertheless recognized that the maintenance of a healthy and viable democracy necessarily entails the maintenance of a healthy and viable community. (p. 179)

Post used this dialectic as the basis for his analysis of free speech rights. The underlying dichotomy also is a useful characterization of the tension between alternative interpretations of the fundamental rights in the electronic media. Traditional common carrier content regulation fits well in the responsive democracy principle; there has been an emphasis on the rights of the individual communicator; the carrier has been given almost no power to influence the content of the message. Broadcast regulation has placed greater emphasis on community; there has been an attempt to balance the rights of the broadcasters and the public. The concept of "public interest, convenience, and necessity" has been interpreted in ways that have required content to be shaped to serve specific community goals. As the media converge, they generate cases that emphasize the conflict among these views of social organization.

The constitutional touchstone for these cases is the First Amendment—a part of the U.S. Constitution, noted for its apparent brevity and clarity, that proves to be illusory as it generates a wide range of complicated and conflicting interpretations. Post characterizes the First Amendment as "a vast Sargasso Sea of drifting and entangled values, theories, rules, exceptions, predictions" (pp. 297–298). In applying the First Amendment to communication technologies, the situation is made more difficult by the fundamental differences between the right of freedom of speech and the right of freedom of the press. Free speech rights have focused on the rights of individual citizens to say what they wish to say. Free press rights have focused on the right of the owners of printing presses (and, by extension, some other communication technologies) to control all the messages distributed by those technologies. The electronic technologies have shifted the ground under the distinction between freedom of speech and freedom of the press, as individuals have sought to "speak" over communication technologies owned by others. This has created a fundamental conflict between two constitutionally protected rights: the rights of those who wish to speak, and the rights of those who own the technology. There has been a delicate and awkward balance between those rights as the courts alternately have emphasized "community" (and, therefore, the rights of individual speakers to be heard) and "responsive democracy" (and, therefore, the rights of owners to control their technology). This balance is being upset by the rapid convergence of technologies. In recent cases, the emphasis has been on the rights of the technology owners to control their property; the balance has shifted away from the rights of individual speakers to be heard, which would result from placing a higher value on community. Although the new media provide the potential for a vastly expanded range of outlets for message creators, the current trend, which places the control of those outlets in the hands of a few corporations, has the potential to turn this virtual cornucopia of expression into a very constrained reality.

THE O'BRIEN TEST AND THE SHAPE OF THE NEW MEDIA

For many years, the courts have struggled with the problem of determining what, if any, government regulations may be imposed on communication media. If the regulation is based on the content of the speech—"distinguish[ing] favored speech from disfavored speech on the basis of the ideas or views expressed" (*Turner v. FCC*, 1994, p. 2459)—the regulation can be upheld only in narrowly defined circumstances. If a regulation is content-neutral, the Supreme Court has indicated that the O'Brien test is the applicable method to determine whether it passes constitutional scrutiny. The Supreme Court's reaffirmation of O'Brien in its 1994 *Turner* (1994) opinion has made this test the key determinant of the constitutionality of communication policies.

In *Turner* (1994) the Court was asked to rule on the constitutionality of the FCC's "Must Carry" rule, which requires cable-TV systems to provide channel capacity to local television stations. The opinion did not resolve that issue. It featured eight different clusters of the Justices across various parts of the plurality opinion and a series of concurring and dissenting opinions. The Must Carry rule was sent back to the lower court for further consideration. Although *Turner*

(1994) did not resolve the status of the Must Carry rule, it did establish the criteria to be used in reviewing content-neutral regulations of communication technologies. *Turner* (1994) restated the three-pronged test from *O'Brien*, which asserted that a content-neutral rule would be upheld only: "[if] it furthers an important or substantial governmental interest; if the governmental interest is unrelated to the suppression of free expression; and if the incidental restriction on alleged First Amendment freedoms is no greater than is essential to the furtherance of that interest." (*Turner*, 1994, p. 2469; citing *Ward v. Rock Against Racism*, 1989).

The Court also emphasized the clarification of the third prong of the test from *Ward v. Rock Against Racism* (1989), which requires that a rule be narrowly tailored: "Narrow tailoring in this context requires, in other words, that the means chosen do not 'burden substantially more speech than is necessary to further the government's legitimate interests' (p. 2469; citing *U.S. v. O'Brien*, 1968).

The word *substantially* is important here because it indicates that the rule can be upheld even without a demonstration that it is the least restrictive regulatory alternative: "So long as the means chosen are not substantially broader than necessary to achieve the government's interest, however, the regulation will not be invalid simply because a court concludes that the government's interests could be adequately served by some less-speech-restrictive alternative." (*Ward v. Rock Against Racism*, 1989, p. 799). Thus, regulations must further important government interests, and must be "narrowly tailored" to impose a minimal burden on the rights of those who are constrained by regulations.

In applying *Ward* (1989), the courts have emphasized the rights of the media owners (particularly cable-TV and telephone companies), and have evaluated regulations in light of the burden they impose on those owners. In these cases, the courts have discounted a community-based view of the burden imposed on free expression in society. The result has been a surprisingly consistent set of district court and appeals court rulings holding that § 533(b), which prohibits telephone companies from entering the video market, is too broadly tailored to pass Constitutional scrutiny. The Supreme Court has granted certiori in *U.S. v. C & P Tel, Co.* (1995), so this consistent picture may change in the future. However, all the current opinions point in the same direction. If that pattern continues, these cases will determine the possible scope for future regulatory policies. Rather than reviewing each of these cases in detail, the next sections of this chapter examine two key dimensions that cut across the cases, and that are fundamental for determining what regulations will pass constitutional scrutiny. Those dimensions are: (a) what, if any, "legitimate government interests" might be protected by regulation? and (b) what, if any, regulations might be sufficiently "narrowly tailored"?

LEGITIMATE GOVERNMENT INTERESTS

While failing to uphold the ban on telephone company entry into home video distribution, the courts have identified a number of government interests that are substantial enough to be the basis for regulations. For the most part, the interests identified have emphasized the government interest in controlling the competitive environment for video distribution. A few opinions have also emphasized a government interest in promoting free speech. To identify these potential government interests, the details of the ways those interests are identified in individual cases are relatively unimportant. The relevant consideration is the range of interests that have been identified.

Fundamental for all these cases is the recognition that the government has a legitimate interest in "protecting the diversity of ownership communications outlets" (*Chesapeake*, 1993, p. 928). In the later opinions, this was presented as a logical extension of *Turner's* (1994) holding that: "assuring that the public has access to a multiplicity of information sources is a governmental purpose of the highest order, for it promotes values central to the First Amendment. Indeed, 'it has long been a basic tenet of national communications policy that the widest possible dissemination of information from diverse and antagonistic sources is essential to the welfare of the public'" (p. 2470). In this context, the courts have viewed the telephone companies as a new source of information that would add to the diversity available to the public.

A second legitimate government interest recognized by the courts was establishing policies that would promote competition in the video industry. Again, *Turner* (1994) established the basic concept that was picked up in later opinions: "the Government's interest in eliminating restraints on fair competition is always substantial, even when the individuals or entities subject to particular regulations are engaged in expressive activity protected by the First Amendment." (p. 2470).Unfortunately for the government's position in these cases, the courts found it difficult to see how keeping the telephone companies out of the video distribution market would promote competition. "Fostering competition in the cable industry is a substantial government interest for First Amendment purposes... It is less clear, however, that § 533(b) fulfills this interest" (*U.S. West*, 1995). In the eyes of the courts, allowing the telephone companies to enter appeared to increase competition, not limit it.

The courts rejected the government's arguments that potential monopoly power of the telephone companies was so great that allowing the telephone companies to enter would result in less competition and a reduction in the diversity of information. They took a skeptical view of the postulated future dangers and followed the *Turner* (1994) dictum: "When the Government defends a regulation on speech as a means to redress past harms or prevent anticipated harms, it must do more than simply 'posit the existence of the disease sought to be cured.' It must demonstrate that the recited harms are real, not merely conjectural." (p. 2470). Although the courts rejected § 533(b) on other grounds, many of the opinions made it clear that they thought the government would have a heavy burden to establish the existence of anticipated harms.

There was one competitive area where the courts generally agreed government had a legitimate interest: preventing cross-subsidization of video ventures from telephone rate revenues. Some opinions went on to note that even an attempt to cross-subsidize would be an anticompetitive act that would raise a legitimate government interest. However, none of the courts accepted § 533 (b) as an appropriate measure to address this interest.

In addition to demonstrating that the alleged harms are real, the government also has the burden of showing "the regulation will in fact alleviate these harms in a direct and material way" (*Turner*, 1994, p. 2470). Because the courts did not accept that the government had demonstrated the existence of real harms in these cases, they did not need to reach the level of evaluating whether the § 533 (b) actually would address those harms. However, many of the courts observed that the FCC had recommended that Congress repeal § 533 (b). In light of that request, they found it implausible that § 533 (b) actually was effective in preventing the alleged harms.

NARROWLY TAILORED

The primary failure of the ban on telephone company video distribution was due to its breadth. "Narrow tailoring" of regulations is required so that they do not excessively limit free speech.

> [A] regulation may [not] burden substantially more speech than is necessary to further the government's legitimate interests. Government may not regulate expression in such a manner that a substantial portion of the burden on speech does not serve to advance its goals... So long as the means chosen are not substantially broader than necessary to achieve the government's interest, however, the regulation will not be invalid simply because a court concludes that the government's interest could be adequately served by some less-speech-restrictive alternative. (*Ward*, 1989, p. 799)

It is not surprising that a complete prohibition on providing video would come to fail this narrow tailoring test: "There is no more draconian approach to solving the problem of potential anti-competitive practices by telephone companies in the cable television industry than a complete bar on their entry into that industry." (*Chesapeake & Potomac*, 1993, p. 927). "Even assuming that the alleged ills prevented by § 533 (b)'s ban are real, the Government has not shown that less restrictive regulations would be ineffective." (*BellSouth*, 1994, p. 1343).

In light of these concerns, several of the courts pointed to more narrowly tailored regulations that might be acceptable. One option that is offered would be to allow the telephone companies to control only some of the channels on their systems: "Congress could simply limit the telephone companies' editorial control over video programming to a fixed percentage of the channels available; the telephone companies would be required to lease the balance of the channels on a common carrier basis to various video programmers, without regard to content." (*Chesapeake & Potomac*, 1994, p. 202). This proposal certainly has some appeal, and is more narrowly tailored than a complete ban.

However, it is possible that it, too, would be held to be an excessive limitation on the speech of the telephone companies. This question is closely related to the question considered in *Turner* (1994)—whether the free speech rights of cable-TV companies are unconstitutionally burdened by requiring that they must carry broadcast stations on some of their channels. If the ultimate disposition of *Turner* (1994) finds that the Must Carry rules are an unconstitutional burden, it is probable that a requirement that telephone companies reserve some of their channels for common carrier access also would be found to be unconstitutional.

A further problem with the common carrier proposal is that it would require the telephone companies to operate in a way that is fundamentally different from the way cable-TV companies operate. Cable-TV companies come to the consumer offering a carefully structured package (or series of packages) of channels. The choice of channels is made to provide the package with the widest possible appeal. The key to structuring the package is the ability of the cable company to control all the channels (at least all the channels other than the over-the-air broadcasters). If the telephone companies were required to sell some of their channels on a common carrier basis, they could lose the ability to develop packages, which are competitively appealing.

Alternatively, the telephone companies could be allowed to form new entities that would offer video programming: "[The proposal of the FCC to allow video programming through a separated subsidiary] would help ensure that the telephone companies do not develop monopolies in the programming market, while allowing them to engage in considerably more protected speech." (*U.S. West*, 1995, p. 41).

For these separations to be successful, cost-separation and accounting rules would need to be developed to maintain the isolation of the video programming business. Although ultimately workable, developing the details of the separation scheme is likely to require a long period of time, which could substantially delay telephone company entry into the video marketplace. Because both cable-TV and telephone companies are faced with a rapidly growing array of alternative delivery systems (direct broadcast system [DBS], etc.), such a delay could make it impossible for the telephone companies to establish themselves in the market. The result could be the same as an outright ban—the video marketplace develops without the participation of the telephone companies.

In deciding the constitutionality of § 533 (b), the courts are not required to consider the viability of more narrowly tailored alternatives. There certainly are alternatives available, but it is less than clear whether any ideas proposed by the courts would actually offer the telephone companies an opportunity to be viable players in the video market.

AMPLE CHANNELS OF COMMUNICATION

Focusing on the First Amendment rights of the telephone companies raises an additional test that any regulation must pass: Does it leave the companies with alternative channels of communication?

> That Section 533(b) is not "narrowly tailored" is not its only infirmity to intermediate scrutiny, however; the provision also does not leave the telephone companies with ample alternative channels for communication. Whether a regulation leaves open ample alternative methods of communication is more than an inquiry as to whether the regulation "completely silences" the speaker. Rather, for a regulation to be constitutional, the ample alternative methods of communication must be sufficiently similar to the method foreclosed by the regulation. (*Chesapeake & Potomac*, 1994, pp. 202–203)

Under this test, it appears likely that any set of "narrowly tailored" regulations that restricted the telephone companies' access to video would fail to provide "ample alternative methods" for communication.

Judge Grady expanded on the lack of alternatives available to Ameritech.

> The fact is, however, that the statute does not leave open ample alternative channels of communication, despite the Government's argument that plaintiffs are free to provide video programming to customers outside their service areas or through independent media outlets. Plaintiffs concede that they could communicate through these channels, at great expense and inconvenience, but the court agrees with plaintiffs that the Government's reasoning here is akin to telling the Chicago Tribune that it may distribute a newspaper everywhere but in Chicago, or that its ability to communicate is not significantly curtailed by its having to publish its news stories in other publications. (*Ameritech v. U.S.*, 1994, p. 736)

The alternative channels that have been proposed for the telephone companies— offering video outside their service areas, and contracting with cable companies to carry their programming—do not appear likely to meet the requirement that they provide "ample" avenues for communication.

REGULATORY OPTIONS AVAILABLE

In addition to the constitutional problems discussed previously, the § 533 (b) ban is also under assault from many other directions. Congressional proposals to allow phone companies to distribute video (and to allow cable-TV companies to offer local phone service) have been close to passage for several years. Thus far, final enactment of these proposals has been delayed because they are just one part of a broad revision of the structure of the communication industries.

These legislative proposals would allow the telephone companies to establish video-distribution systems through separate subsidiaries, with the states and the Federal Communications Commission (FCC) charged with guarding against cross-subsidization. In addition, they would allow the creation of "video platforms," which would allow others to lease channel capacity on the systems. If the ultimate disposition of *Turner* (1994) overturns the Must Carry rules for cable-TV it is probable that these leased-channel requirements will face a similar constitutional challenge.

The FCC also has been actively revising its procedures in light of the continuing series of opinions rejecting § 533 (b). In May 1995, it announced that it would permit waivers to allow telephone company construction of video systems.

> In response to decisions of the Fourth and Ninth Circuits, we conclude that under Section 613(b)(4) we have the legal authority to grant waivers allowing telephone companies to provide video programming in their telephone service areas on video dialtone networks. We adopt that construction of the waiver provision because it is fully consistent with the language of the statute and Section 613(b)'s underlying policy, and because waiving the restriction in that manner obviates the constitutional infirmities identified by the courts of appeals. (Federal Communications Commission, 1995, p. 1216).

While proclaiming the availability of such waivers, the FCC did not provide details on how the waivers would operate.

> In particular, we do not decide whether telephone companies may provide video programming over video dialtone networks rather than as traditional cable operators. In addition, if a telephone company is permitted to provide video programming on a video dialtone system, we are not here deciding whether that telephone company should be regulated under Title II or Title VI of the Communications Act. Nor do we decide the conditions under which telephone companies may be granted waivers to provide traditional cable service in their telephone service areas. (p. 1215)

The FCC was not entirely open ended in its description of the conditions it envisioned. It indicated that such features as common carriage of signals and a ban on cross-subsidy would be part of the final regulations.

> Video dialtone necessarily includes a common carriage element, and we have previously concluded that a telephone company may not allocate substantially all of its capacity to a single "anchor programmer." Accordingly, a telephone company providing video dialtone service is not allowed to occupy all of the channels provided by the system, but has a common carrier obligation to make capacity available to others. The common carrier aspect of video dialtone service promotes both competitive and free speech interests by making room for more than one speaker. Second, our current video dialtone rules contain provisions intended to ensure that telephone companies providing video programming directly to subscribers do not discriminate in favor of their affiliated programmers and do not subsidize video programming operations with rates collected from their provision of monopoly telephone services. These restrictions are intended to promote the underlying purpose of Section 613(b) by fostering fair competition in the multi-channel video programming market. (p. 1219)

Here, the FCC has tried to carefully tie these requirements to the constitutionally approved governmental purposes of promoting fair competition.

INITIAL FIRST AMENDMENT RIGHTS OF COMMON CARRIERS: THE DIAL-A-PORN PRECEDENT

Although most of the recent discussion has focused on the First Amendment rights being given to telephone companies for providing video services, it is important to recognize that similar rights are being granted in other areas. Telephone companies traditionally were regulated as common carriers that distribute the messages of others without responsibility for the content of those messages. In recent years, the ground has shifted by establishing the right of the telephone companies to determine the content of some of the audio messages they carry.

The emergence of this right has been an unanticipated consequence of the establishment of the 976 services. On these services, the telephone company bills subscribers on behalf of the information providers who are called. In addition to the many services providing traditional forms of information, a number of information providers sought profits through sexually oriented messages. Collectively, these message services came to be know as "dial-a-porn." Some subscribers found the availability of these services highly objectionable, and there were a number of attempts at regulation by various states and the federal government.

The Supreme Court in *Sable* (1989) resoundingly rejected attempts at governmental prohibition of indecent messages. The justices were unanimous in the finding that adults had a right to receive indecent telephone messages, and that a complete governmental ban, even one that was adopted with the constitutionally permissible goal of protecting minors, would infringe this adult reception right: "Because the statute's denial of adult access to telephone messages which are indecent but not obscene far exceeds that which is necessary to limit the access of minors to such messages, we hold that the ban does not survive constitutional scrutiny." (*Sable v. FCC*, 1989, p. 131). For the telephone companies, this ruling was a mixed blessing. Although they were pleased that a government regulation of their industry was rejected, they were faced with the problem of billing for indecent messages that were offensive to some very vocal customers. Under a traditional common carrier analysis, the telephone companies would have been required to provide service to the indecent message providers.

Fortunately for the companies, they also were gaining discretionary powers to reject these messages. Justice Scalia's concurring opinion makes his views explicit: "I note that while we hold the Constitution prevents Congress from banning indecent speech in this fashion, we do not hold that the Constitution requires public utilities to carry it." (*Sable v. FCC*, 1989, p. 131). Although the government was prohibited from enacting a ban, the companies were not required to carry the messages. Most operating companies moved quickly to disassociate themselves from billing for the dial-a-porn services.

This development was consistent with the earlier resolution of *Carlin v. Mountain States* (1987). Carlin operated a 976 service, which the appeals court characterized as "salacious" and "offering smut." Mountain Bell terminated Carlin's service a few days after receiving a letter from an Arizona county deputy attorney threatening prosecution if it continued to carry Carlin. In the following week, the Mountain Bell board adopted a new policy that denied 976 service for all "adult entertainment." The appeals court found the state's threat to be unconstitutional, but upheld the company's ban.

> [T]he initial termination of Carlin's service was unconstitutional state action. It does not follow, however, that Mountain Bell may never thereafter decide independently to exclude Carlin's messages from its 976 network. It only follows that the *state* may never *induce* Mountain Bell to do so. The question is whether state action also inhered in Mountain Bell's decision to adopt a policy excluding all "adult entertainment" from the 976 network. We hold that it did not. (*Carlin v. Mountain States*, 1987, pp. 1296-1297)

Carlin's attempt to have its service restored to allow it to continue to distribute its message was rejected. The opinion indicated that even a public utility could refuse to carry some messages: "Mountain Bell may exercise some business judgment about what messages, even lawful ones, it will carry." (*Carlin v. Mountain States*, 1987, p. 1294). The opinion does not go on to discuss the extent to which "business judgment" may be used as the criteria for selecting messages. The opinion makes clear that future arrangements for service are entirely up to the parties involved.

> Mountain Bell is now free to once again extend its 976 service to Carlin. Our decision substantially immunizes Mountain Bell from state pressure to do otherwise. Should Mountain Bell not wish to extend its 976 service to Carlin, it is also free to do that. Our decision modifies its public utility status to permit this action. Mountain Bell and Carlin may contract, or not contract, as they wish. (*Carlin v. Mountain States*, 1987, p. 1297)

In short, the government is effectively precluded from exercising control, and the telephone company is free to carry or reject content as it wishes. It is important to note that obscene content can be prohibited by law, although indecent content may not be. The Supreme Court has denied certiori to the attempt to declare the revised Helms' Amendment ban on obscene phone services to be unconstitutional. Regardless of whether it is desirable for the telephone companies to have these powers, the Court appears to believe it has a First Amendment right to grant or deny service based on message content. Although the limits of this right have not been determined, the opinions suggest these rights are likely to expand further. The similar precedents in cable-TV, suggest that this transition is well underway.

The 1992 Cable Television Consumer Protection Act required cable programmers to identify all indecent programmers so that cable operators could place them on a channel that would be available only to those requesting it. In addition, cable operators were required to block obscene material on any access channel. The D.C. Court of Appeals upheld these regulations, stating: "[The] segregation and blocking requirements satisfy the least restrictive means test; do not impermissibly single out leased access programming for regulation; do not constitute a prior restraint on speech; and are not, because of the definition of indecency, unconstitutionally vague." (*Alliance v. FCC*, 1995, p. 67). This opinion points to a future in which cable operators and telephone video suppliers will have both the power and the obligation to control the content on their leased access channels. Barron (1993) develops the implications of the dial-a-porn cases for other media. "[A] transition from a common carrier model to a *de facto* model is exactly what happened in the telco experience with 'dial-a-porn.' We must be careful that such an unexamined transition does not occur once telcos take us into a new information age." (p. 404).

In light of the similarity of the results in the telephone company First Amendment rights cases and the cable operator First Amendment rights cases—increasingly giving content control to system operators—it appears this transition is here. Thus, it is important to begin to structure regulations for this new environment.

REINTRODUCING THE CONCEPT OF COMMUNITY

In these cases—establishing the right of the telephone companies to provide video services—much of the emphasis has been on the rights of the companies to control their property. As with the owners of printing presses, they are being given the right to use their facilities to communicate as they wish. Unfortunately, most of the court opinions have tended to ignore the direct link between property rights and speech rights. "To regulate property *is* to regulate speech. The Founders understood that, which is why they protected both, equally. To watch the modern Court try to determine which is dominant, whether the regulation of property or the regulation of speech, is to watch a morality play without direction because [it is] without foundation. And that will not change until we get back to basics." (Pilon, 1994, p. 63).

However, these basics need to include a redressed balance between property rights and the community's interest in diversified forms of information.

Within the parameters defined by the growing set of cases emphasizing First Amendment rights for system operators, there is also the possibility that those rights can be realized in ways that provide for diverse exchanges of views. For those who are concerned with promoting maximum diversity of content, there are two potentially viable alternatives available. One would be to encourage the development of multiple systems of delivery—to try to create a environment in which diverse programming could provide a competitive advantage. The second would be to establish a regulatory framework in which companies would voluntarily give up their control of some content in exchange for other benefits.

Although the recent cases have emphasized the First Amendment rights of the system operators, it is important to bear in mind that there are a number of other categories of participants who have First Amendment rights that have yet to be asserted. For instance, program producers have some First Amendment rights to develop messages as they see fit. Also, members of the receiving audience have some rights to have access to the content they prefer. It is impossible to maximize the rights of all three of these groups simultaneously. Granting extended rights to any of them can only be done at the expense of the rights of the others. If the system owners have complete control over the content of their systems, then both producers and the audience are left without direct influence. If producers are given the right to demand that their content be carried (as the broadcasters are seeking to do with the Must Carry requirement), then system owners and the audience are deprived of other content they might have selected. If the audience is given complete control of content, then the rights of both the system owners and the producers are violated. The resolution of these issues will continue to be a dynamic balancing of the rights of all of these groups.

As the rights of one group are expanded, the rights of the other groups necessarily are restricted. A recent proposal in the *Harvard Law Review* suggests that this be resolved by seeking to promote "viewpoint diversity." This approach would extend broadcast-type regulation to the new media under a revised scarcity rationale.

> [The] frequency scarcity rationale and the regulatory edifice built upon it were aimed at minimizing *viewpoint* scarcity. The apparently limited number of broadcast channels threatened to restrict the number of viewpoints that would be expressed. This danger, in turn, imperiled the First Amendment rights of *viewers and listeners* to receive the broad range of ideas necessary to create a well-functioning marketplace of ideas. ("The Message in the Medium," 1994, p. 1074)

The key to resolving the First Amendment dilemmas of the new media is to return to more fundamental principles: "The transient nature of communications technologies requires that the evolution of the First Amendment be grounded in the underlying interests that the First Amendment seeks to protect. Focusing on such guideposts makes the prospect of a rapid change in the First Amendment less threatening and that hope for a regulatory regime that advocates free speech more realizable." ("The Message in the Medium," 1994, p. 1098).

This sort of fundamental regrounding of the First Amendment analysis certainly is an appropriate course of action. However, it appears unlikely to be followed, in light of the technological and property emphasis the courts and the litigants are choosing in the emerging cases, which actually are determining the new environment.

Rather than seeking a clarified understanding of First Amendment principles, the Court appears to be struggling to apply an increasingly complex set of technology- and property-based precedents to identify the First Amendment rights of the corporate participants in the cases before it. From this perspective, the cable operators and telephone companies are seen to be analogous with publishers, with the result that they are found to have extensive First Amendment rights to control the entire range of content over their facilities.

In a separate opinion in *Turner* (1994) Justice O'Connor (joined by Justices Scalia, Ginsburg, and Thomas) unambiguously sees the control to be in the hands of the cable operator: "The question is not whether there will be control over who gets to speak over cable—the question is who will have this control. Under the FCC's view, the answer is Congress, acting within relatively broad limits. Under my view the answer is the cable operator." (*Turner*, 1994, p. 2481). O'Connor recognizes that there are dangers in giving this much power to the system operator, and goes on to suggest that Congress may choose to encourage other new media as a means of providing a more diverse selection of messages. The Court is just one vote short of a majority that would recognize sweeping First Amendment powers of the operators. Because the other justices also find that the system operators have extensive First Amendment rights, it appears clear that the Court will continue to extend those rights in the future. It is likely that the Court will give similar emphasis to the property rights of the telephone companies when it addresses those issues.

Although the courts are giving greatly expanded rights to control content to the system owners, there are other cases that suggest that this content control may not be absolute. In *PruneYard* (1980), the Court upheld the power of California to require shopping centers to allow solicitation. In addition to the content-neutral character of the requirement, the court noted:

> [T]he shopping center by choice of its owner is not limited to the personal use of appellants. It is instead a business establishment that is open to the public to come and go as they please. The views expressed by members of the public in passing out pamphlets or seeking signatures for a petition thus will not likely be identified with those of the owner... [The owners] can expressly disavow any connection with the message by posting signs in the area where the speakers or handbillers stand. (p. 87)

Perhaps this could be extended to video distribution systems by requiring a form of access on some channels. Operators may have a First Amendment right to choose to structure systems that are not "open to the public to come and go as they please." But the economics of the business of video distribution, like the business of shopping centers, may strongly favor systems that provide some open access. Even if system owners are given complete First Amendment control over content, there are significant practical, legal, and business reasons they might prefer an environment where some of that control was shifted to the hands of others.

From a practical standpoint, the logistic demands of supervising the content of 500 channels of video on a broadband distribution system are overwhelming. As numerous broadcasters have found, it is difficult for management to exercise complete control of the content for even a single radio station. Twenty-four hours of content requires 24 hours of supervision. With multiple channels, the management problems become much more complex. If the world of switched video develops, with its unlimited number of potential video sources, it will be impossible to maintain management control over all of them. Faced with the enormous time and resources required to make content decisions, system operators may be interested in seeking regulatory structures that allow them to transfer some of those decisions, and thus some of the accountability and responsibility, to others.

The legal obligations imposed on system operators who have complete control over content will be substantial. They also will be responsible for that content, and can be sued based on the content of the programming they have selected. Under the well-established tort principle of suing the people with the money, the telephone companies with their vast assets will be tempting targets for future actions. Dee's (1994) discussion of tort litigation on behalf of children who allegedly were traumatized by indecent dial-a-porn messages provides a glimpse of some of the litigation that may develop in video programming. In this environment, the system operators might be very interested in regulatory schemata that would assign some content control to others in exchange for protection from litigation.

Finally, the fundamental business interests of the system operators may cause them to seek diverse programming. They can be successful only if the public actually uses their systems. That will require a wide range of content to attract as much use as possible. Unfortunately for the companies, the content that attracts some users will be offensive to others. The companies will need to explore schemata where individual users can block out the content they find offensive, even though the system carries that content to others who wish to receive it. Rather than relying on system-wide content controls, or even on channel-by-channel controls, system operators may find it to their advantage to allow subscribers to have program-by-program control of access. This would result in an environment where the system operator would want to distribute as many different types of programming as possible, while the user would have complete control over the range of options that would enter the home. Preserving the concept of community with its associated diversity of messages may turn out to be a wise business decision, even before the pendulum of First Amendment jurisprudence again moves in that direction.

REFERENCES

Alliance v. FCC, 53 F.3d 105, U.S. App. LEXIS 13857 (1995).

Ameritech Corp. v. United States, 867 F. Supp 721, 736 (N.D. Ill. 1994).

Barron, J. A. (1993). The telco, the common carrier model and the First Amendment—the "dial-a-porn" precedent. *Rutgers Computer & Technology Law Journal, 19,* 371–404.

BellSouth Corp. v. United States, 868 F. Supp 1335, 1343 (N.D. Ala. 1994).

Carlin v. Mountain States, 827 F.2d 1291, 1292 (9th Cir. 1987); cert. denied, 485 U.S. 1029 (1988).

Chesapeake & Potomac Tel. Co. v. United States 42 F.3d 181, 202 (4th Cir. 1994).

Chesapeake & Potomac Tel. Co. v. United States, 830 F. Supp. 909 (E.D. Va. 1993).

Dee, J. (1994) To avoid charges of indecency, please hang up now: An analysis of legislation and litigation involving dial-a-porn. (1995). *Communications and the Law, 16,* 3–28.

Federal Communications Commission. Telephone Company–Cable Television Cross-Ownership Rule. Third Report and Order 77 Rad. Reg. 2d (P&F) 1216 (1995).

The Message in the medium: The First Amendment on the information superhighway (1994). *Harvard Law Review, 107*, 1062–1098.

Pilon, R. (1994). A modest proposal on "Must-Carry," the 1992 Cable Act and regulation generally: Go back to basics. *Hastings Comm/Ent L.J., 17*, 41–63.

Post, R.C. (1995). *Constitutional domains: Democracy, community, management.* Cambridge, MA: Harvard University Press.

PruneYard Shopping Center v. Michael Robins, 447 U.S. 74 (1980).

Sable v. FCC, 492 U.S. 115 (1989).

Turner v. FCC, 114 S.Ct. 2445 (1994).

U.S. v. C & P Tel. Co., cert. granted, 63 U.S.L.W. 3906 (26 June 1995).

U.S. v. O'Brien, 391 U.S. 367 (1968).

U.S. West, Inc. v. United States, 1994 U.S. App. LEXIS 39121, *28 (9th Cir. 1995).

Ward v. Rock Against Racism, 491 U.S. 781, 799 (1989).

13

The Economics of Scholarly Publications and the Information Superhighway

Roger G. Noll
The Brookings Institution and Stanford University

Electronic technology offers many interesting opportunities for lowering the costs, increasing the variety, and speeding the dissemination of scholarly publications. Unfortunately, electronic publication also presents some new problems and aggravates others. Specifically, because electronic duplication is much cheaper than duplication of hard copies, in the absence of effective methods for preventing unauthorized use, electronic distribution can reduce the economic viability of scholarly publications and lead to less, rather than more, effective dissemination of new knowledge. The purpose of this chapter is briefly to describe the economics of scholarly publication and how it may be affected by electronic distribution.[1]

The structure of the essay is as follows. First, a basic model of demand and supply in markets for scholarly publication is developed, including the entry process for new publishers of books and journals. Next, the effect of illegal duplication on this market is explored. Then, the economic and technical characteristics of electronic publication are examined, including their likely effect on the market that has developed using traditional technology. Finally, some conclusions are offered about the likely effect of new technology on scholarly publication, the dissemination of research, and university libraries.

THE SOURCES OF DEMAND

Scholarly publications serve several distinct uses, each of which in some measure creates demand for scholarly books and journals. One function of scholarly pub-

---

1. The economics of scholarly publications is a special case of the economics of information (see e.g., Owen, 1975; Noll, Peck & McGowan, 1973). For an explanation of the economics of information products for noneconomists, see Noll (1993).

lications is the dissemination of information, ranging from fundamental new concepts to cleverly organized aggregations of old data. Scholars frequently minimize the importance of this use of scholarly publications because, by the time scholarly works are published, they are often outdated as representations of the current research frontier. Research scholars tend to rely primarily on preprints, seminars, and conferences for information about the latest advancements in scholarly knowledge.

Nevertheless, dissemination through publication is important for practitioners and students. In scholarly fields with extensive practical uses, such as science, engineering, finance, and law, scholarly publications are relied on extensively in industry and government to guide applications. For these users, whether a publication contains results at today's frontier of knowledge is not valued as highly as among research scholars. Scholarly publications are also useful in educating students. As in more applied research, most education is not focused on the current frontier of research, even at the graduate level. Consequently, scholarly publications are intensively used for advanced undergraduate and graduate education. Furthermore, even if scholars rarely use publications to learn about new developments in their research specialties, they do use them to keep abreast of developments in other fields.

Scholarly publications also function as tools for evaluating researchers. The advantage of scholarly publications is that the prepublication review process and subsequent citations (as an indicator of impact) are important sources of information about scholarly standing. In addition, the hierarchy of journals also generates signals about the quality of scholarly work that facilitates dissemination and evaluation. The scholarly publication process is used as a screening mechanism to enable researchers to allocate their searching and reading time more efficiently.

Both of these factors cause scholarly publications to have an economic value. Scholars derive value from a more efficient process for evaluating and disseminating their work; universities derive value from the improvements in their evaluation process that are made possible by scholarly publications; and students, other scholars, and participants in the private economy derive benefit from the dissemination functions of scholarly publications. But all of these sources of value have public goods properties (i.e., to an individual, the benefits of the process do not depend fully on whether a subscription is actually purchased). For example, scholars and universities derive the benefits of evaluation and prestige that flow from the dissemination of a publication, regardless of whether either buy it. One employee of a high-technology firm can make use of an engineering journal that was bought by another employee or by a nearby library.

Because of this "public good" feature of scholarly publication, each beneficiary has some incentive to "free ride"—not to pay a "fair share" of the costs of scholarly publications. As a result, the effective demand for scholarly journals—the actual response in the market to the introduction of a publication of a given quality that is offered at a given price—is likely to be an underestimate of the true economic benefits, as measured by the aggregate maximum willingness to pay.

The preceding argument supports the tentative conclusion that scholarly publications are likely to be undersupplied—a conclusion that is at variance with the commonly held opinion that many scholarly publications have little or no value. (The conclusion is tentative because assessing the efficiency of scholarly publications requires consideration of costs, which are examined in the next section.) The contrary argument runs something like this. First, most scholarly publications are simply not read by more than a handful of people, and are never again cited by anyone other than the author, who presumably did not need to have the work published to know about and cite it.[2] Second, many people express the belief that scholars place too much emphasis on research and not enough on teaching; university service; being good citizens, parents, and spouses; and so on. Third, because scholars do not bear the full cost of publication, they can be expected to overuse it.

These arguments are incomplete. Some of the purposes of scholarly publication do not depend on attracting readers. In particular, the signal "published, not cited" is different from "published, cited" and "failed to be published." In addition, the likely impact of a scholarly work cannot be perfectly anticipated in the review process. If peer assessments of submitted manuscripts are subject to significant random error, the expected number of readers and citations can be positive at the time of publication, even if, *ex post*, the median number of readers and citations is zero. If scholars knew in advance which of their works were unimportant, or if peer review were perfect, a reduction in publication could be accomplished with zero social cost. However, with imperfect *ex ante* assessments, if one observed a circumstance in which all publications were widely read and cited, the number of publications would assuredly be too few.

Finally, whereas scholars do not bear the full cost of publication (although in the sciences they frequently bear a part of it), publishers do, and find it worthwhile. Moreover, the cost of publication represents a relatively small fraction of the social cost of producing and disseminating research. Scholars, their employers, and other sponsors of research bear a large fraction of the total costs of scholarly work, in terms of the opportunity costs of the time spent undertaking the research and writing the manuscript.

Of course, a comprehensive assessment of the extent of scholarly publications requires consideration of costs and the mechanics of the market for scholarly publications. But with respect to the efficiency of the scope of scholarly publications, two important facts are important. First, many specialized journals that are infre-

---

2. For example, the top 10% of scholarly journals in economics account for 88% of all citations, and 74% of economics journals have never received a single citation in any of the top 11 journals (see Pencavel, 1992).

quently cited pass a market test, especially in technical fields. Moreover, in technical fields, a significant share of this market is accounted for by libraries and employees of government and industrial laboratories. Second, judging from such factors as citations and appearances on reading lists, the best articles in secondary journals and books from secondary publishers have substantially more impact than the worst publications from the most prestigious sources. This supports the view that *ex ante* assessments of the significance of scholarly works are subject to substantial uncertainty. Hence, the notion that a reduction in the number of scholarly publications would eliminate only the least valuable material is clearly optimistic.

THE SUPPLY OF SCHOLARLY PUBLICATIONS

Like all information products, the supply of scholarly publications can be separated conceptually into two distinct components: (a) the production of the basic information product (in this case, the edited, formatted, and composited publication), and (b) the duplication and dissemination of this product (in traditional technology, printing and shipping). In addition, the supplier undertakes marketing costs.

The first component is called the *first-copy costs*. First-copy costs include the work of the author in producing a manuscript and the work of editorial personnel in evaluating and preparing it for publication. For scholarly publications, the first-copy costs include all of the research by the scholar, plus refereeing and editing.

The most important feature of information products is that the magnitude of first-copy costs is independent of the second-stage costs. That is, the cost of the content of the publication is independent of how many copies are produced. To a rough approximation, then, the costs of an information product can be decomposed into a linear function:

$$C = F + mq, \tag{1}$$

where C is total cost, F is the first-copy cost, m is the cost of producing and distributing one more copy of the published product (the marginal cost of a copy), and q is the number of copies that are produced and distributed. This simple representation has only two important complications that need to be noted, but that are subsequently ignored to simplify the argument.

First, for most publications, the number of copies produced exceeds the number of copies distributed. The cause of excess production is uncertainty about final demand. For reasons explained later, the price of a publication is usually substantially higher than the marginal cost of production. Consequently, the publisher will print one more copy if the probability of selling it is reasonably high, but less than one. Hence, the optimal printing run is usually higher than the expected number of sales. This phenomenon adds complexity to the conceptual model of the supply of information products, but does not really affect the core conclusions.

Second, marketing costs do not fit neatly into either first-copy costs or duplication and dissemination costs. Indeed, marketing costs have elements of both. Some aspects of marketing are like first-copy costs: preparing the promotional material. Other aspects are more directly related to the quantity sold, in that the intensity of advertising is positively related to the number of copies sold. Hence, both F and m in Equation (1) contain marketing elements.

The most important implication for all information products that are derived from this generic characterization of costs is this: For a publication to be economically viable, the price of a copy must exceed the marginal cost of production. That is, total revenues must generally not fall below m in order for a publisher to survive in the market. Hence, if P is the price of the publication, it must be true that:

$$P \geq F/q + m. \tag{2}$$

This basic formula is somewhat more complicated if the publication sells advertising, but not greatly so. Typically, advertising rates depend on the size of the ad, the location of the ad in the publication, and the circulation of the publication. Hence, the price on the left-hand side of Equation (2) can be decomposed into two elements: the subscription price and the advertising prices for each size and location of ad. Thus, if there are k types of ads, P can be rewritten as:

$$P = P_s + P_1 Q_1 + \dots + P_k Q_k, \tag{3}$$

where the first term on the right-hand side of Equation (3) represents the price charged to consumers, the remaining Ps represent the prices per subscriber for various types of ads, and the Qs represent the number of ads that are sold. This complication does not alter the basic fact that the revenues per copy from all sources must exceed the marginal cost of producing one more unit.

One important reality about all products that have this kind of cost function (including broadcasting and all other information products) is that the most efficient pricing system is not consistent with the financial viability of the product for a private producer.[3] Neither subscriptions nor advertising can be priced at marginal cost; together they must generate enough revenues to cover the first-copy costs (F in Equations [1] and [2]) if the publication is to be financially viable. By setting price above marginal cost, however, some potential subscribers will be excluded who otherwise could be provided an incremental copy of the publication at a price that exceeded its marginal cost. The necessity to exclude such subscribers to cover

---

3. The following argument closely parallels the economics of pay television (see Spence & Owen, 1977).

total costs constitutes a fundamental efficiency problem for all private markets for information. This problem can only be solved by subsidy.

Another important implication of this general cost function is that the price of a publication, all else equal, is likely to increase if circulation falls. A drop in circulation is likely to cause a drop in advertising revenues, but a less than proportionate drop in costs, because the first-copy costs must still be covered. Hence, subscription revenues per subscriber must cover a higher fraction of total costs. Although the actual pricing effect depends on the precise way in which demand for the publication declines, the most common case is that a fall in demand leaves the publication with buyers who value it most highly and have less price-sensitive (less elastic) demand, in which case the price will increase (or, in some cases, the material will not be published).

Scholarly publications differ from commercial publications in a number of important ways. Most important, scholarly publications are not expected to cover the parts of the first-copy costs that represent the author's effort. Research and writing costs are paid by the author's employer and, in some cases, a grantor or even the author. Although book publishers usually pay royalties, on average these are small compared with the first-copy costs of research and manuscript preparation, and are not the primary source of the author's motivation to publish. In addition, some scholarly publishers are subsidized by universities, professional associations, and foundations. Hence, the costs that they must cover may be less than the fixed and marginal costs of publication and distribution.

In many cases, scholarly publications impose charges on the author. One form is a submission fee; another is a page charge. In addition, some scholarly publications require that manuscripts be submitted in machine-readable form, or that tables and figures be submitted in a form that is ready for photo-offset printing. These requirements are equivalent to a publication charge, in that they shift some of the first-copy costs to the author. Thus, the simple formula in Equation (2) requires further amendment for some journals to take into account these other forms of revenues.

Entry and exit of publications is determined by whether the maximum feasible incremental revenues from a publication enable it to cover its total costs. Imagine that all scholarly publications were controlled by a monopolist.[4] In considering whether to publish one more book or add one more scholarly journal, the monopolist would inquire whether the increment to revenues would exceed the increment to costs arising from the additional publication. In making this calculation, the monopolist would consider whether the new publication caused some customers to switch from another publication. The monopolist would "count" only those revenues of a new

4. The discussion that follows was first developed with respect to competition in broadcasting by Peter O. Steiner. For a recent comprehensive statement of this approach to understanding competition in the media, see Owen and Wildman (1992).

publication that were generated by a net increase in purchases; it would ignore revenues that represented a substitution of the new for the old. By contrast, a second firm contemplating a new publication would count as incremental any revenue that shifted from another publisher. The decline in revenues, and hence profits, suffered by the first publisher would be irrelevant to the assessment of the desirability of an additional publication by the second publisher. This basic logic is the basis for a conceptual understanding of the process that determines the number of scholarly books and journals that are published. To some extent, publications within a given discipline or field of research are substitutes. An increase in total publications will reduce the sales and average revenues of publications in the field.

The markets for journals are especially interesting in their dynamics. Initially, journals were few in number and generally covered entire disciplines, reflecting the relatively small demand that supported the market. As time progressed, more scholars sought publication, and rejection rates increased. The established journals experienced increased circulation and revenues. In practice, the initial monopoly suppliers—usually professional associations and university presses—could have responded by expanding the number of journals, but typically they did not. Instead, they increased rejection rates to improve the quality of their publications. In some cases, the initial publishers used the increased profits from journals to subsidize other activities of professional associations. Consequently, excess demand for new publications grew, broadening the process of scholars' assessment and allowing easier and broader dissemination of work that was thought to have value, but was rejected by the established journals.

Initially, entry took the form of more general purpose journals, but the reputation and first-in advantage of the established general disciplinary journals prevented entrants from achieving circulation equivalent to their competitors. Eventually, entry began to take the form of specialized journals, largely because of the character of demand. Individuals and corporate laboratories, contemplating the choice between the fifth or sixth best general purpose journal (with relatively few articles in the special field most relevant to the subscriber) and a specialized journal in the relevant field, would tend to pick the latter. Likewise, authors, although preferring the prestige of a top-ranking general purpose journal, sought to target their other publications in journals that would be read by the most people in their fields, rather than skimmed or skipped by a general audience. Hence, for journals of lesser quality, specialization came to be dominant. The phenomenon has come to be called *journal proliferation*, referring to the tendency for new journals to come into the market to serve narrower and narrower specialties as research volume increases.

The effect of journal entry is to reduce average circulation among older journals, because each journal derives some of its subscribers from individuals and institutions that cancel other subscriptions to free funds to buy the newcomer. As circulation declines, the average cost of older journals increases. Moreover, in the case of all but the general circulation journals, the remaining demand for the older journals typically becomes less price-sensitive (less elastic). Each specialized journal increasingly relies on a subscriber base of narrow specialists for whom the journal represents the best way to reach other similar specialists among journals lacking the prestige of the leading general journals. Hence, as more journals enter, prices for established journals generally increase.

By the late 1980s, the process had created literally thousands of journals with relatively low circulation, numbering a few hundred subscribers.[5] For example, imagine a journal that does not require an especially fancy publication process (e.g., no color photos or complex figures). If the journal publishes about 150 pages per issue, it will face a marginal cost of printing and distribution in the range of $2–$4 per issue, or, for four issues per year, of $8–$16. Its first-copy costs will be in the range of $50,000–$100,000, depending on whether the editor and referees are paid, and on the extent of copyediting and proofreading. If it has a circulation of 400 (roughly the median for scholarly journals), the first-copy costs per subscriber will be between $125–$250. Hence, its subscription price must be far in excess of marginal costs for the journal to survive. In this example, a fall in circulation has little effect on costs, and so mainly causes an increase in the minimum viable price for the journal.

In addition, because the gap between minimum viable price and marginal cost is so large, the market for low-circulation journals is likely to be highly inefficient, with many potential subscribers excluded. Publishers deal with this problem, to some extent, by attempting to discriminate between individual and institutional subscribers. Frequently, individual subscriptions are as little as 10% of the institutional price, and frequently are in the range of one fourth to one third. But even so, individual prices are usually substantially above marginal costs. In addition, the high institutional price causes libraries to be far smaller than would be socially optimal. Of course, for publications in science and engineering, this inefficiency ripples throughout the entire economy. This is so because education, applied research and development, and direct diffusion to the production of goods and services will proceed at a slower rate than otherwise would be the case.

The phenomenon of declining average circulation and rising prices has placed increasing pressure on general purpose libraries, such as libraries at universities and colleges. The combined effect of more journals and, due to declining average circulation, higher prices led to rapidly rising costs of maintaining a comprehensive library of scholarly journals. Many commentators interpreted these developments as representing monopolistic practices of publishers.[6] First, the price increases

---

5. The basic facts about prices, costs, and circulation reported here are based on my research in collaboration with Lisa L. Cameron and W. Edward Steinmueller. A summary of the information about prices and circulation is contained in Noll and Steinmueller (1992). A more detailed econometric study is in preparation. In addition, see Marks, Nielson, Petersen, and Wagner (1991).

6. For example, see Association of Research Libraries (1989), Selsky (1988), Tuttle (1990), and Hamaker and Tagler (1988).

were said to represent monopoly pricing. Second, journal proliferation was explained by the practice of publishers inducing undeserved subscriptions by paying leading scholars to lend their names as editors to unimportant new journals.

In practice, neither explanation is a fully consistent explanation. Journals may have market power, but there is no reason to believe that it has increased. In any case, declining average circulation of journals is sufficient to explain the pricing phenomenon. Likewise, the identification of a journal with a famous scholar is undoubtedly beneficial for its initial marketing. But if it did not contain material that was useful for dissemination purposes (and did not generate citations), it would not generate permanent subscribers. Industry would not subscribe and universities would not face resistance to canceling subscriptions if neither faculty nor students wanted access to it. This identification with a famous scholar might shift high-quality articles among journals, but it cannot explain the increase in the total number of financially viable publications. The latter is due to the increase in the number of useful publications, which in turn reflects the persistent rise in the number of researchers and real expenditures on scholarly research.

Libraries have responded to price increases and journal proliferation by cutting back on subscriptions and placing greater reliance on interlibrary loan. For any single library, this strategy makes sense if a journal is not frequently used. For a few dollars, a request for a journal can be satisfied by interlibrary loan, and a faculty member or student can legally photocopy the desired article. If such a request occurs only a few times a year, the library can save several hundreds of dollars in subscription costs (plus a few more in journal storage costs) for an outlay of tens of dollars in the interlibrary loan process.

Unfortunately, for all libraries taken together, this process probably does not reduce total costs, and may even increase them. The reason is that only a small part of journal costs are related to the marginal cost of adding one more subscriber. The savings to the publisher is not the hundreds of dollars for the subscription, but the few dollars in printing and distributing one more copy. Hence, the total effect of interlibrary loan on all libraries is to raise their aggregate costs by increasing journal prices and causing libraries to incur the added costs of interlibrary loan (including additional photocopying). For publications that are used with some frequency, the savings from lower distribution costs are not likely to offset the additional costs of interlibrary loan. Consequently, the perfectly rational strategy for each library is collectively irrational: It reduces the quality of library services and has no significant effect on costs (and may increase them).

ILLEGAL COPYING

The economics of illegal copying of scholarly publications fits nicely into the supply and demand model presented in the preceding discussion. The cost of illegal duplication is the sum of the direct cost of producing a copy (c), the fine for being caught (f) times the probability of detection (p), and the psychic disutility from engaging in illegal behavior (d).[7] Hence, a person will engage in illegal copying rather than acquiring a personal copy if:

$$P > c + pf + d. \tag{4}$$

The extent of illegal duplication will change if any of these factors changes.[8] Assume for purposes of discussion that two of these factors are fixed: the fine for illegal copying and the personal distaste for illegal activity. This assumption enables one to focus on the effects of events that might affect the other elements of Equation (4).

Clearly, the trend in rising journal prices can be expected to increase illegal copying. If Equation (4) is not satisfied because P is slightly too low, an increase in P will cause a change in behavior—from subscribing to copying illegally. Hence, the fall in average circulation from journal proliferation eventually leads to a further fall in average circulation because the proportion of users who gain access through illegal copying increases. Because libraries are more likely than individuals to be caught and fined if they copy illegally, price increases are likely to cause the proportion of subscriptions accounted for by institutions to increase. This process further increases the cost of general purpose research libraries, and intensifies the gap between the individual and collective rationality of sharing arrangements between libraries.

In similar fashion, a fall in the real price or an increase in the quality of copying also increases illegal duplication. The most important technological events that have taken place to date are the invention and subsequent improvement of photocopying. Prior to photocopies, illegal duplication required a form of publication. Although illegal publication could be of somewhat lower quality, the main saving was to avoid the editorial and marketing costs and, in the case of books, royalties to authors. The illegal duplicator still had to pay for preparing the material for pub-

---

7. There is good evidence that the distaste for extensive illegal duplication is significant. For example, some popular and profitable computer software is distributed as "shareware." That is, the software is given away for free on computer bulletin boards or in published articles, or is sold at roughly the cost of duplication on diskettes. Users are then asked to pay the license fee (usually in the range of $25–$100) only if, after testing the program, they decide to use it permanently. Because enforcement of the intellectual property rights in shareware is essentially impossible, these payments are voluntary, yet a large number of people pay them (see Takeyama, 1994a).

8. Some copiers would never subscribe because they value the publication at less than P, but more than the cost of illegal copying. These people do not harm the publisher by illegally copying the material. In the case of scholarly publications, some illegal copying actually can enhance the value of the publication to the publisher if it increases citation, and hence the reputation and visibility, of the publication (Takeyama, 1994b). This effect can be important for journals, where cumulative past citations are likely to affect the demand for subscriptions.

lication, and then printing and distributing it. Because the duplicator did not offer much in cost saving, illegal duplication of scholarly works was rare.

The introduction of photocopying vastly reduced the costs of duplication by eliminating the first-copy costs of the process. It also vastly increased the problem of detecting illegal copying, so reduced pf as well as c in Equation (4). Subsequently, falling real costs of photocopying, due to technological progress, further increased the incentive to engage in illegal copying. Because the cost of photocopying includes the time of the person who does the copying, the vastly faster photocopiers introduced in the 1980s made illegal duplication more attractive. Thus, technological progress in photocopying constitutes a separate factor that drives down average circulation and increases subscription prices for libraries, much in the same fashion as does interlibrary loan.

Duplication of publications by even the most efficient photocopiers is usually inefficient. This inefficiency depends on the relationship between m and c. In general, photocopying costs (including the time of the photocopier) are greater than the marginal cost of printing and distributing one more copy. But for the vast majority of scholarly journals, the reader usually does not want to read everything. If the reader places little or no value on most of the journal's contents, copying a few desired items can be cheaper than paying even the marginal distribution cost for a copy of the journal. Consider the previous example of a journal that costs, say, $3 per issue to print and distribute (not counting first-copy costs). Suppose that the cost of photocopying is $.10 per page for the photocopier, plus 10 seconds per page in the time of the person doing the duplicating. If this person earns $15 per hour, this adds about 4 cents per page for copying. Then photocopying an entire 120-page journal would cost $10 for photocopying, plus $5 in employee costs, or $15, which is far more than the marginal cost of one more issue of the journal. But photocopying one 6-page article costs $.60 in photocopying, plus one minute of employee time (another $.25). Thus, if a scholar wants to read one short article per issue, it is cheaper in total social costs to photocopy the article than to print one more copy of the entire journal.

Of course, publishers are unlikely to make some arrangement with a scholar to permit this kind of efficient duplication. A publisher would have no incentive to make such an arrangement unless it could recover some of the first-copy costs from users of this type. In attempting to set a price above the actual costs of duplication, the publisher would drive the cost of legal duplication too high in relation to the cost of illegal duplication, at least for users with a small disutility for illegal duplication. In addition, the cost of negotiating separate deals to allow copying could well exceed the cost saving from user photocopying. Hence, in a manner parallel to the fundamental inefficiency of average cost pricing for all information products, efficient user duplication is likely to be impossible to achieve legally.

THE CONSEQUENCES OF NEW INFORMATION TECHNOLOGY

A common belief among library administrators and other potential users of electronic media for the distribution of scholarly publications is that the information superhighway offers prospects for solving the problems of escalating journal publication costs. In the framework of the analysis presented here, this belief is almost certainly incorrect.

The only source of cost savings from electronic distribution of scholarly publication is in the cost of storing and accessing publications. Electronic distribution will still need to cover the same first-copy costs associated with reviewing manuscripts and editing the final product. Whereas computer technology has reduced first-copy costs, these savings do not depend on how the publications is distributed. A recent study (Bowen, 1995) estimated that storage costs are about $2 per year per book or journal, and that reshelving hard copies after use costs another $1. When added to the few dollars of printing and distributing hard copies, these savings are substantial compared with institutional subscription prices only for journals that are rather extensively used. Most likely, for most journals and books, the savings arising from electronic publication, at best, would equal one year's rise in the average cost of publications.

In addition, electronic publication reduces the cost of illegal duplication because it avoids the cost of photocopying. If a reader wants a hard copy of an article, printing an electronic file and photocopying hard copy have about the same costs. However, the costs of storing illegal copies electronically is low, so these costs can be avoided if a hard copy is not needed.

An interesting issue arises with respect to the ease of detection of electronic copying. At present, methods for protecting software are not very secure, but they will become more so in the future; detection might be easier than it is now for either photocopying or pirating software. But once a user has access to a legal electronic copy, willful duplication for others is almost impossible to detect unless the copy is stored on the hardware of a computer that is connected to a public network. Legal access can be gained by a single legal electronic subscription, or by a hard copy subscription and a scanner. Thus, electronic duplication is potentially as important an innovation as photocopying in terms of its effects on illegal copying. If so, it will cause legal subscriptions to decline and, consequently, institutional subscription prices to increase. Hence, electronic publication will add to, not subtract from, the financial problems of libraries and the inefficiency of the distribution process among institutions, where illegal acquisition and copying is easier to detect and is subject to harsher penalties.

In addition, electronic publishing will certainly increase the range of feasible methods of producing scholarly product. For example, it will vastly reduce the cost of including data sets, full motion video, and nonlinear textual material. All of these are likely to be valuable additions from the perspective of disseminating scholarly information. But they are also likely to increase the first-copy costs of publications by making publications longer and more complex. If so, electronic publication will increase, not reduce, the financial pressures on libraries and the inefficiencies arising from the pricing structure of scholarly publications.

Even without illegal duplication, electronic distribution threatens publishers. Many scholars already post preliminary versions of their publications on the World Wide Web (WWW). Potential readers can make use of the screening and classification features of the publication process by simply consulting the advertising material of the publishers, listing recent books and journal contents. After using this material, readers can then bypass the publisher by downloading the manuscript from the author's Web site, and perfectly legally bypass the publication process—and the payment of a share of the editorial costs of the publisher.

A common reaction to these problems is simply that groups of scholars can band together to publish "approved" lists of manuscripts that are available on the Internet, thereby avoiding the need for journals altogether. Although such groups may well emerge, whether informally or formally, they will not change the fundamental problem. This concept amounts to a proposal to shift the first-copy costs to groups of volunteer scholars. It will succeed only if the editorial inputs of publishers (such as copyediting and marketing) are of little value, and if the inputs of scholarly editors (such as providing in-depth refereeing and detailed suggestions for improvement) are either of little value or will be supplied without remuneration. None of these requirements seems particularly plausible; if they were, journals would already have eliminated the services of little value and/or moved to rely exclusively on volunteer services.[9]

CONCLUSIONS

The preceding analysis leads to several important conclusions. The first is that, in the absence of relatively secure methods for preventing unauthorized access and duplication, publishers are not likely to embrace electronic publication of existing scholarly products because it offers little prospective cost saving, but a considerable threat in terms of increased illegal duplication. The ripest prospects for electronic distribution are either new types of publications that make use of the new technical capabilities of digital communication, or historical documents that have little future market value. Even so, journals are especially in trouble because scholars place working papers on the Internet and make them freely available. Internet access is likely to be used, to some degree, as a substitute for the purchase

---

9. Margaret Jane Radin, in commenting on an earlier version of this chapter, pointed out that the vast majority of law reviews are edited almost entirely by uncompensated students. She also observed that, judging from the results, this example is not one that other disciplines should rush to follow.

of final publications by business and libraries, which will further erode average circulation of journals.

Second, the financial problems of major libraries are unlikely to be ameliorated by electronic technology. Libraries can capture savings in storage and circulation costs by shifting to electronic copies, but these are small for all but the most heavily circulated publications. Working in the opposite direction, electronic publication is likely to increase the institutional price for legal access because few individuals will subscribe as more engage in illegal copying.

Third, because electronic publication will lead to more illegal duplication and innovations that take advantage of the quality-enhancing capabilities of electronic media, the ratio of first-copy costs to total costs of scholarly publications will rise. As a result, the prices of publications will also increase. Hence, the inefficiencies of the scholarly publication market—in particular, undercirculation to institutional subscribers—will get worse.

Fourth, from a societal perspective, electronic publication and more widespread circulation should be encouraged. The best means for accomplishing this objective is subsidization of the fixed cost of publication, so that publishers would need to rely on subscriptions only to recover distribution costs. Unfortunately, subsidies are not easy to implement: How would the subsidizing institution determine whether a publication deserved a subsidy? Most likely, any system for allocating subsidies would depend on some sort of peer review process to evaluate publications, and would therefore be likely to erect barriers to entry of new journals.

In the context of the underlying market for scholarly publications, electronic technology raises many interesting and formidable problems. Although electronic media offer exciting new prospects for expanding the form of publication, and can offer some real cost savings, they may aggravate the economic problems that have already arisen in scholarly publication.

ACKNOWLEDGMENTS

The author gratefully acknowledges the useful comments on a previous draft from Margaret Jane Radin and Gregory Rosston, and research support from the Brown Center for Education Policy, the Markle Foundation, and the Mellon Foundation.

REFERENCES

Association of Research Libraries. (1989). *Of making books there is no end.* Washington, DC: Association of Research Libraries.

Bowen, W. G. (1995). JSTOR and the economics of scholarly communication. *The Economics of Information in the Networked Environment*, Meredith A. Butler and Bruce R. JKingma, editors. Washington: Association of Research Libraries, 1996, 22–34.

Hamaker, C., & Tagler, J. (1988). The least reading for the smallest number at the highest price. *American Libraries, 18*, 764.

Marks, K. E., Nielsen, S. P., Petersen, H. C., & Wagner, P. E. (1991). Longitudinal study of scientific journal prices in a research library. *College and Research Libraries*, 125–138.

Noll, R. G. (1993). The economics of information: A user's guide. *The knowledge economy: Annual review of the institute for information studies*. Wye, MD: Aspen Institute, 25–52.

Noll, R. G., Peck, M. J., & McGowan, J. J. (1973). *Economic aspects of television regulation*. Washington, DC: Brookings Institute.

Noll, R. G., & Steinmueller, W. E. (1992). An economic analysis of scientific journal prices: Preliminary results. *Serials Review, 18*, 32–37.

Owen, B. M. (1975). *Economics and freedom of expression*. Lexington, MA: Ballinger.

Owen, B. M., & Wildman, S. S. (1992). *Video economics*. Cambridge, MA: Harvard University Press.

Pencavel, J. (1992). *Comments on "The Journals of Economics" by George J. Stigler, Stephen M. Stigler, and Claire Friedland*. Presented at centennial anniversary conference on Scholarly Communication, University of Chicago, April.

Selsky, D. (1988). Librarians face continued price increases of books and periodicals. *Library Journal, 118*, 28.

Spence, A. M., & Owen, B. M. (1977). Television programming, monopolistic competition, and welfare. *Quarterly Journal of Economics, 91*, 103–26.

Takeyama, L. (1994a). The welfare implications of unauthorized reproduction of intellectual property in the presence of demand network externalities. *Journal of Industrial Economics, 4*, 155–66.

Takeyama, L. (1994b). The Shareware Industry: Some stylized facts and estimates of rates of return. *Economics of Innovation and New Technology, 3*, 161–74.

Tuttle, M. (1990). The pricing of British journals for the North American market. *Library Resources and Technical Services, 30*, 72–78.

On the Duration of Copyright Protection for Digital Information

Yuehong Yuan
Carnegie Mellon University

Stephen F. Roehrig
Carnegie Mellon University

This chapter is a study of the implications of new digital technology on the duration of copyright protection. Digital, networking, and satellite technologies are causing dramatic changes in information markets. These changes include (a) decreased copying and distribution costs, (b) intensified competition ("every reader is a potential publisher"), (c) increased importance of first-copy development as a result of copying and distribution cost reduction, and (d) globalization of information markets.

In the face of these sweeping changes, copyright protection remains the fundamental legal underpinning in the provision and consumption of information. It encourages information production by protecting the rights of publishers to recover development costs and make profits. It also encourages information usage by limiting the duration of copyright protection.

Copyright laws have existed, at least in the West, for more than 200 years. They have been adapting to changes caused by technologies. However, the current challenge from digital technology is no doubt a major one. The implications for copyright protection can be profound. There may be a need to change some provisions in copyright law; there may be a need for new technical methods to implement copyright protection.

There has been an increasing number of studies on new technical methods for enforcing copyright law in the digital world. Techniques proposed range from prevention, such as password protection, to detection, such as hidden "watermarks" in digital documents (e.g., Brassil, Low, Maxemchuk, & Gorman, 1994). Al-

247

though further development of technical methods are still very much needed, it is fairly safe to assume that new protection technology will eventually be there to solve the problems caused by technology.

There have been extensive reviews on copyright laws. The U.S., Canadian, European, and Japanese governments have performed or sponsored reviews of copyright laws of their respective countries (e.g., Advisory Council on Information Highway of Canada, 1995; Lehman, 1995). In the current Congress of the United States, there are several bills pending to revise the 1976 Copyright Act (S. 1122, S. 483, S. 1284, H.R. 989, R.H. 2441). There is also heated public and academic discussion—including publications, copyright conferences or conference sessions, and active mailing list discussions—on copyright law.

This chapter contributes to the discussion on copyright and new digital technology by quantitatively modeling and analyzing the change of duration of copyright that might be warranted by changes in the information market caused by new digital technology. Duration is a key factor in a copyright system. Copyright has gone through a long trend toward lengthening its duration. The first statute of copyright, the Statute of Anne, enacted in England in 1709, granted authors an original term of 14 years from the date of publication, plus a second term of 14 years should the author be living at the expiration of the first term (Guinan, 1963). The first federal statute on copyright in the United States in 1790 followed the 28-year maximum tradition of the Statute of Anne. The duration was lengthened to 42 years in 1831 (28-year original term plus 14-year second term), and was lengthened to 56 years in 1909 (28-year original term plus 28-year second).

The 1976 Copyright Act of the United States made the duration the author's or the last surviving author's life plus 50 years; for anonymous works, pseudonymous works, and works made for hire, the duration is 75 years from publication or 100 from creation, whichever expires first.

This trend of lengthening of copyright duration is continuing. The Copyright Term Extension Act of 1995, introduced in both the U.S. Senate and the House of Representatives (S. 483 and H.R. 989), extends duration of copyright for 20 more years. The new proposed durations are author's or the last surviving author's life plus 70 years; for anonymous works, pseudonymous works, and works made for hire, duration is extended to 95 years from publication or 100 from creation, whichever expires first.

There is a danger that this trend of lengthening is brought about by a configuration of political forces not reflective of the economic rationale behind copyright protection, as originally sanctioned in the Copyright Clause in the U.S. Constitution. This particular configuration of political forces may exist because of the uneven distribution of the effects of a longer duration of protection. Generally, publishers have a lot to win with a longer duration of copyright, whereas the possible adverse effect of a longer duration is dispersed throughout society, with no single section being particularly hurt. Thus, the political forces representing publishers may be more mobilized than those representing the general public.

Given this background, an assessment of a proposed extension of copyright protection would be very useful. However, to assess if a proposed extension of duration is in the right direction, or if the extension is enough, two prerequisite tasks need to be accomplished: (a) an assessment of the appropriateness of the original duration under old technological and market conditions, and (b) changes of duration commensurate with technological and market changes determined. This chapter tries to do only the second. Thus, this chapter must be complemented by an evaluation of the current duration to assess a proposed extension of copyright protection.

The duration of copyright has been modeled by Braunstein (1980), who, under the assumption of monopoly, studied copyright protection and optimal duration of protection for software and databases. However, pure monopoly control is rare even in the traditional information market. Protected information often faces competition from substitutes: different, but similar, information products from other creators. In the traditional market, one may resort to the control of printing machines and distribution channels, which are costly to obtain and operate, to suppress competition. In a digital world, the cost of reproduction and distribution is decreased substantially. This decrease in reproduction and distribution cost may inevitably loosen the grasp of monopoly based on the control of reproduction machines and distribution channels. Thus, the digital information market can be expected to be more competitive.

Furthermore, a copyright system grants a monopoly only in reproduction and distribution. Competitors are free to develop similar products independently or through reverse engineering. The key feature that makes a copyright system preferable to public development is that such a system facilitates competition in first-copy development, which is absent in public development. Thus, it is essential to consider competition in modeling duration of copyright protection.

A model with competition is formulated in the next section, followed by a discussion that solves the model under simplifying assumptions. Then, the implications of the solution for the duration of copyright in the face of new market changes are discussed. The effect of suboptimal protection on social welfare and information availability is presented thereafter. Finally, some final remarks are offered.

A GENERAL MODEL FOR OPTIMAL DURATION UNDER COMPETITION

The goal of this chapter is to study implications on copyright duration of changes in various market conditions. Hence, a fairly general model of duration is formulated. A general model is one of monopolistic competition, where suppliers have monopoly on their products, but compete with suppliers of similar products. Perfect competition and monopoly can be seen as special cases of monopolistic competition. Monopolistic competition is also more reflective of real market situations, because pure monopoly and perfect competition are both unlikely.

Assume there are n publishers (where $n \geq 1$) developing first-copy information products and selling copies of them. For simplicity, assume first-copy products of all publishers are developed simultaneously at Time 0. Copies are made and sold thereafter. Each publisher incurs a major cost in first-copy development at Time 0 and incurs costs of reproduction and distribution over time.

Within the duration of copyright protection, each publisher is a monopoly in reproducing and distributing copies of products developed by itself. Thus, it can charge a copy price higher than the marginal cost of reproducing and distributing the copy. This "higher-than-marginal-cost" pricing makes it possible for publishers to recover first-copy development costs and make profits. At the same time, it causes losses in consumer welfare, because consumer demand with a marginal utility higher than the marginal cost of copying and distribution, but lower than the monopoly price charged by each publisher, will not be satisfied. Thus, the utility exceeding marginal copying and distribution cost is foregone.

Beyond the duration of protection, anyone can make and sell copies of the information products. Prices per copy of products will thus be driven down to their respective marginal cost of copying and distribution. In this period, publishers cannot earn revenue beyond copying and distribution costs to recover first-copy development costs or make profits. However, all demand with marginal utility equal to or higher than the marginal costs of copying and distribution will be satisfied. No consumer welfare achievable through the first-copy products will be lost.

Thus, the duration of protection is critical to publishers' ability to earn a revenue beyond copying and distribution costs. This determines how much they can spend to develop first-copy products, and how many first-copy products will be developed. It also affects how much consumer demand for given first-copy products will be fulfilled.

The social welfare brought by publication generally equals the sum of consumers' valuation of each copy bought, net reproduction and distribution costs, and net first-copy development costs. A long duration of protection would reduce the number of copies that reach consumers. A short duration of protection would reduce the incentive for publishers to develop first-copy products in the first place. One task of copyright legislation is to choose an optimal duration of protection that maximizes social welfare, thus balancing the incentive for publishers to develop first-copy products and the fulfillment of consumer welfare.

In general, the demand for each publisher's products depends on the quality, variety, and prices of its information products and those of others. Competition between publishers gives them incentive to improve quality, increase variety, and reduce price. However, because this chapter investigates the effect of intensified competition and decreasing copying and distribution cost, it is assumed that all information products of all publishers are of the same category and quality. "The same category" means that differences between software and movies are not considered. There is only one category of product—only movies—for example. "The same quality" means that all movies are "equally good," such that any potential referee cannot say any movie is better than others, even though they are all different movies having different plots, characters, and so on.

Quality difference and product differentiation are two distinct concepts. Quality difference is not considered here. But all first-copy products are different from each other. Thus, the variety of information products is reflected in the number of first-copy products. If the number of first-copy products is bigger, the total demand for copies of the products will generally be higher. The increase in total demand with the number of first-copy products will depend on the degree of product differentiation.

With the prior simplification, the demand for information products of any publisher will depend on the number of first-copy products and the price per unit copy charged by this publisher, and the numbers of first-copy products and prices per unit copy of other publishers.

The remainder of the chapter uses the following notations:

S_i: the number of first-copy products of publisher i, $i=1, 2,..., $ n;

S_{-i}: the vector of numbers of first-copy products of other publishers except publisher i;

p_{it}: the price per unit copy of publisher i at time t;

p_{-it}: the vector of prices per unit copy of other publishers, except publisher i, at time t;

$D_{it}(s_i, p_{it}, s_{-i}, p_{-it}, t)$: the rate of demand for products of publisher i at time t, which increases in s_i and p_{-it} and decreases in p_{it} and s_{-i};

$c_i(s_i)$: the cost of first-copy development of s_i;

b: the cost of copying and distributing unit copy of information, uniform over all information products of all publishers; and

T: the duration of copyright protection.

The problem of optimal duration of copyright can be formulated, in general, as choosing the duration of protection T to maximize social welfare L—that is,

Maximize

$$
L = \sum_{i=1}^{n} \int_{0}^{\infty} \left(\int_{b}^{\infty} D_{it}(s_i, p, s_{-i}, p_{-it}, t)\, dp \right) e^{-\gamma t} dt - \sum_{i=1}^{n} c_i(s_i)
$$

$$
- \left(\sum_{i=1}^{n} \int_{0}^{T} \left(\int_{b}^{p^*} D_{it}(s_i, p, s_{-it}, p_{-i}, t)\, dp \right) e^{-\gamma t} dt \right.
$$

$$
\left. - \sum_{i=1}^{n} \int_{0}^{T} D_{it}(s_i, p_{it}{}^*, p_{-it}, s_{-it}, t)\,(p_{it}{}^* - b)\, e^{-\gamma t} dt \right)
\tag{1}
$$

subject to

$$
\frac{\partial}{\partial p_{it}} [D_{it}(s_i, p_{it}, p_{-it}, s_{-i}, t)\,(p_{it} - b)] = 0, \quad i=1, 2, ...,n
\tag{2}
$$

and,

$$
\frac{\partial}{\partial s_i} \left(\int_{0}^{T} D_{it}(s_i, p_{it}{}^*, p_{-it}, s_{-i}, t)\,(p_{it}{}^* - b)\, e^{-\gamma t} dt - c_i(s_i) \right) = 0, \quad i=1, 2, ..., n
\tag{3}
$$

where γ is the discount rate taken to be the same for publishers and consumers, and p^*_{it} is the selling price determined in Equation (2).

The first term in Equation (1) is a measure of the maximum possible consumer surplus (accounting for copying and distribution cost b) obtainable from s_i ($i=1,...,$ n). The second term is the cost of developing s_i. The remaining terms within the large parentheses represent the deadweight loss due to copyright protection of duration T, which is the unfulfilled consumer surplus less publishers' revenue above copying and distribution cost during T. Some deadweight loss of this type is inevitable due to the following constraints.

The first set of constraints (2) represents the decision of each publisher in choosing price p_{it} in selling copies of its information within the copyright period. This price maximizes selling revenue, net copying, and distribution cost at each Time t. The second set of constraints (3) describes how each publisher chooses the number of its first-copy products to maximize the total profit from Time 0 to T, accounting for all costs of copying, distribution, and development. Both constraints are first-order conditions for each publisher to maximize profit.

The prior formulation accommodates various levels of competition. When n goes to infinity, perfect competition is achieved; when $n = 1$, it reduces to a model of monopoly. However, this general formulation is hardly tractable. Further assumptions are needed to solve the model and gain insight from it.

SOLUTION UNDER SPECIAL ASSUMPTIONS

In this section, a special form for the demand function is assumed. Let

$$D_{it} = D_0 \frac{s_i}{S} f(S) D(p) g(t) \qquad (4)$$

where $S \equiv s_1 + s_2 + ... + s_n$ is the total of first-copy products of all publishers. In this formulation, D_0 represents the general level of consumer demand for information. D_0 for a large country should be greater than the D_0 for a small country. D_0 for the whole world should be greater than D_0 for a single country. D_0 also reflects the general level of quality or value of information. That is, higher quality, or more valuable, information is represented as higher demand for information at a given price.

The function $f(S)$ represents the effect of the number of first-copy products on the total demand. In general, the bigger S is, the more choices consumers have, and so the bigger the demand. Thus, $f(S)$ will usually increase in S. However, one information product may partially substitute for another. Thus, although adding one more product should create some additional demand, it may also draw some of its demand from other products. Thus, the increase of $f(S)$ in S will typically be less than proportional to S (i.e., sublinear). In the extreme case, when $f(S)$ is constant, different information products are perfect substitutes; when $f(S)$ is linear in S, one information product is "completely different" from others; there is no substitution. The change of $f(S)$ with S represents the degree of product differentiation. In rarer cases, if information products are complements, $f(S)$ is superlinear.

The function $g(t)$ represents the change in demand over time. The demand function can be scaled by adjusting D_0, such that $g(0)=1$. In most cases, the demand for an information product will decrease over time, so that $g(t)$ is a decreasing function. For some information products (e.g., a daily news program), $g(t)$ would decrease to zero quickly. For other information products (e.g., the works of Shakespeare), demand lasts for a long time, perhaps even to infinity. Thus, the time it takes for $g(t)$ to decrease to zero incorporates some notion of the economic life of the product. For example, the economic life of a product can be measured as the time it takes for the demand decrease to zero [or to a third of its initial demand or some other similar measure, depending on the specific form of $g(t)$]. In rarer cases, demand for an information product may rise over time. Finally, $D(p)$ represents the price effect on demand. Generally, $D(p)$ decreases in p.

The main assumption implicit in Equation (4) is the separation of the factors of time, price, and first-copy size. No interaction is assumed among the three factors. Each factor affects demand independently. Another assumption implicit in Equation (4) is that the share of total consumer demand of one firm is the same as its share of the first-copy products of all publishers. When a publisher increases its number of first-copy products, it both increases its share of total demand and increases the total demand for all products. While one publisher's growing number of first-copy products draws away demand from products of other publishers, the others publishers may benefit from the increase in total demand, although the former effect may be dominant. One example where this might be true is the expansion to critical mass of digital information drawing people away from paper-based information. Thus, all digital information publishers may benefit.

The third assumption is that publishers do not engage in price competition. This, along with the fact that $g(t)$ has been made the only component of the demand dependent on t, will allow us to write $p_{it}=p$ for all i and t. This assumption is consistent with the assumption of the absence of quality difference between products within publisher and across publishers, and is consistent with the intention here to focus on competition in first-copy development. A plausible institutional environment where price competition may not be excessive is a commercial digital library (e.g., Yuan, Roehrig, & Sirbu, 1995), which is trusted by publishers to store and sell all their products. If there were no quality difference between products, all products would be priced uniformly.

Acknowledging that situations contrary to the prior assumptions exist, it is nevertheless expected that they reflect, to a reasonable degree, the competition in first-copy development, and can yield insights about copyright duration of digital information independent of the specifics of the formulation. With the demand function given previously, the problem of optimal duration becomes choosing T to maximize

$$L = D_0 f(S) G(\infty) \int_b^\infty D(p)\,dp - \sum_{i=1}^n c_i(s_i)$$

$$-D_0 f(S) G(T) \left(\int_b^{p^*} D(p)\,dp - \left(p^*-b\right)D\left(p^*\right) \right) \tag{5}$$

subject to

$$\frac{\partial}{\partial p}[(p-b)D_{it}(p)] = 0 \tag{6}$$

and

$$D_0(p^*-b)D(p^*)G(T)\left[\frac{f(S)}{S} + \frac{s_i f'(S)}{S} - \frac{s_i}{s^2}f(S)\right] - c_i'(s_i) = 0, \quad i=1,2,...,n \tag{7}$$

where $G(T) \equiv \int_0^T g(t) e^{-\gamma t} dt$ and $G(\infty) \equiv \int_0^\infty g(t) e^{-\gamma t} dt$, and p^* is the value of p satisfying Equation (6). More specific functional forms can be chosen for $f(S)$ and $c_i(s_i)$. Assume

$$f(S) = S^\alpha \tag{8}$$

with $0 \leq \alpha \leq 1$ representing the degree of product differentiation. For $\alpha = 0$, one has perfect substitution, while $\alpha = 1$ indicates no substitution between information products. Further, assume that all n publishers have identical cost functions

$$c_i(s_i) = cs^\rho \quad , \quad i=1, 2,..., n \tag{9}$$

where c represents the general level of first-copy development cost and $\rho > 1$ represents decreasing return in first-copy development. From the symmetry of the n publishers, one has $s_1 = s_2 = ... = s_n = s$. Using Equations (8) and (9), from the second constraint in Equation (7), one can get

$$s = \left[\frac{\alpha + n - 1}{c\rho} n^{\alpha - 2} D_0\left(p^* - b\right) D\left(p^*\right) G(T) \right]^{\frac{1}{\rho - \alpha}} \tag{10}$$

Substituting Equations (8) and (9) into the objective in Equation (5), one has

$$L = \left(D_0 G(\infty) \int_b^\infty D(p)\, dp \right) n^\alpha s^\alpha - ncs^\rho - D_0\left(\int_b^{p^*} D(p)\, dp - \left(p^* - b\right) D\left(p^*\right) \right) n^\alpha G(T) s^\alpha \tag{11}$$

To maximize L, the first-order condition $\frac{\partial L}{\partial T} = 0$ is examined:

$$\left(D_0 G(\infty) \int_b^\infty D(p)\, dp \right) n^\alpha \alpha s^{\alpha - 1} s' - nc\rho s^{\rho - 1} s'$$

$$- D_0\left(\int_b^{p^*} D(p)\, dp - \left(p^* - b\right) D\left(p^*\right) \right) n^\alpha \left[G'(T) s^\alpha + G(T) \alpha s^{\alpha - 1} s' \right] = 0 \tag{12}$$

Dividing both sides by $s^{\alpha-1}s'$, one has

$$\left(D_0 G(\infty)\int_b^\infty D(p)\,dp\right)n^\alpha \alpha - nc\rho s^{\rho-\alpha} - D_0\left(\int_b^{p^*} D(p)\,dp - \left(p^*-b\right)D\left(p^*\right)\right)n^\alpha\left[G(T)\tfrac{s}{s'}+G(T)\alpha\right] = 0 \quad (13)$$

Taking the derivative of Equation (10) and rearranging terms, one has:

$$G'(T)\frac{s}{s'} = (\rho-\alpha)G(T) \tag{14}$$

Substituting Equations (10) and (14) into Equation (13), one gets:

$$\left(D_0 G(\infty)\int_b^\infty D(p)\,dp\right)n^\alpha \alpha - \left[(\alpha+n-1)n^{\alpha-1}D_0\left(p^*-b\right)D\left(p^*\right)G(T)\right]$$

$$-D_0\left(\int_b^{p^*} D(p)\,dp - \left(p^*-b\right)D\left(p^*\right)\right)n^\alpha G(T)\rho = 0 \tag{15}$$

Note that the general demand level parameter D_0 cancels out. Solving for $G(T)$, one has

$$G(T) = \cfrac{\alpha G(\infty)\displaystyle\int_b^\infty D(p)\,dp}{\left(1-\dfrac{1-\alpha}{n}\right)\left(p^*-b\right)D\left(p^*\right)+\rho\left(\displaystyle\int_b^{p^*} D(p)\,dp - \left(p^*-b\right)D\left(p^*\right)\right)} \tag{16}$$

In Equation (16), p^* is obtained from Equation (6). The term $\int_b^\infty D(p)\,dp$ in the numerator represents the rate of maximum possible consumer surplus obtainable from the first-copy information developed; the term $(p^* - b)D(p^*)$ in the denominator reflects the rate of the monopoly revenue—net copying and distribution cost,—of the publishers within the duration of copyright protection. The term $\int_b^{p^*} D(p)\,dp - \left(p^*-b\right)D\left(p^*\right)$ reflects the rate of deadweight loss resulting from copyright protection. One can solve for T by first computing the value of the right-hand side of Equation (16), then finding T such that $G(T)$ equals this value. Because $g(t)$ is non-negative, $G(T)$ is non-decreasing in T. Thus, analyzing the change of $G(T)$, one can analyze the change of optimal duration T of copyright protection with changes in other parameters.

INSIGHTS FROM THE SOLUTION

Changes in the various parameters in Equation (16) represent changes in the information market. The effects of changes in parameters on the optimal duration of copyright protection are analyzed next. Specifically, the effects of intensified competition, decreasing cost of copying and distribution, globalization of markets, and some other factors are reviewed.

Shorter Duration Under Intensified Competition

The development of the Internet, and networking technology in general, may reduce the capital cost of publishing. Setting up a Web site for commercial publishing will be technologically mature and economic viable in the near future. It will also be much easier then setting up a printing press and associated distribution channels. Thus, the barriers to entrance to publishing in a networked digital environment are likely much lower than in the traditional paper environment. One can expect more publishers and more intensified competition in publishing in a networked digital environment. Also, because reproducing and distributing information become less of a challenge, competition will center more on first-copy development and less on reproduction and distribution.

In the model described here, this intensified competition is represented by a larger n—the number of publishers. Recall that, in most cases, the product differentiation parameter α is between 0–1. From Equation (16), $G(T)$ decreases with n. Thus, in general, the model shows that T is smaller when n becomes bigger. Under a monopoly, $n=1$, the optimal duration reaches its maximum, with

$$G(T) = \frac{\alpha G(\infty) \int_b^\infty D(p)\, dp}{\alpha \left(p^* - b \right) D\left(p^* \right) + \rho \left(\int_b^{p^*} D(p)\, dp - \left(p^* - b \right) D\left(p^* \right) \right)} \tag{17}$$

When n goes to infinity, the optimal duration is minimum, with

$$G(T) = \frac{\alpha G(\infty) \int\limits_{b}^{\infty} D(p)\, dp}{\left(p^* - b\right)D\left(p^*\right) + \rho \left[\int\limits_{b}^{p^*} D(p)\, dp - \left(p^* - b\right)D\left(p^*\right)\right]}$$

Giving more specific forms to function $D(p)$ and $g(t)$ and assigning reasonable values to the parameters, one can solve for the optimal duration T. Let

$$g(t) = 1 - \frac{t}{\beta}, \text{ for } 0 \le t \le \beta,$$

and $g(t) = 0$ otherwise. Thus, demand for copies of an information product is assumed to fall off linearly until the end of its economic life at Time β. Thereafter, its demand becomes zero. For reasons of algebraic convenience (but reflecting the inverse relationship between demand and price), further assume an exponential demand function:

$$D(p) = e^{-\lambda p}$$

As a concrete example, let $\left[D_0 \ b \ \alpha \ \beta \ c \ \rho \ \lambda \ \gamma\right] = \left[100000 \ 1 \ 0.5 \ 50 \ 10000 \ 1.2 \ 0.2 \ 0.05\right]$. Note that these values are taken only to resemble a general pattern of relative magnitudes of these parameters, and should not be considered as actual values for these parameters. Thus, the optimal duration obtained with these parameters should not be taken as a suggested duration to any copyright legislation, but as an arbitrary base necessary to find the direction of change of optimal duration if there are changes in the parameters.

With these parameter values, one may solve for p^* from Equation (6), then T from Equation (16) for any given number of n. For $n = 1$ (monopoly), one gets $T = 45$ years, whereas when n grows large, T asymptotically approaches 16 years. The general decrease of optimal duration of copyright protection with the number of competing publishers is shown in Figure 14.1. One feature of Figure 14.1 is that, as the number of competing publishers increases, the optimal duration converges quite quickly to the level for perfect competition. For example, optimal duration is 17 years when $n = 5$, and it is close to 16 years when $n = 10$. This quick convergence might reduce the impact of intensification of competition on optimal duration. Although digital technology may create substantially more competing publishers, the effect on optimal duration of copyright protec-

Figure 14.1
Effect of competition on optimal duration

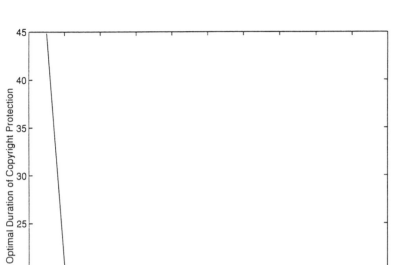

tion may be marginal if the number of competitors in paper-based publishing is already beyond a relatively small threshold. However, the quick convergence in Figure 14.1 may be the result of the specific functional forms and parameter values assumed here. Slower convergence with other functional forms and parameter values is certainly possible.

Thus, in general, the duration of copyright protection should be shorter for digital information to the extent that barriers to entrance to digital publication are lower, allowing for more publishers. Thus, if a publisher enjoys a natural monopoly, a longer protection for copyright is needed to maximize social welfare. This may sound counterintuitive because the publisher enjoying natural monopoly presumably already makes excess profit. But an information product is different from a traditional product. A producer with a monopoly on a non-information product can indeed often make excessive profit. However, if the product of a monopoly publisher is allowed to be freely copied, the publisher may still not be able to recover its development cost. Thus, without copyright protection, a publisher cannot make excessive profit by simply enjoying natural monopoly in first-copy development. Copyright protection is needed for any publisher to recover costs and stay in operation.

The reason that a natural monopolist needs longer protection, and thus be granted a longer legal monopoly in selling its product, may be the following. First-copy information is less available when a natural monopoly exists. Thus, at the margin, first-copy information is more valuable to society. Thus, society will be willing to incur a larger deadweight loss by granting a longer legal monopoly to the publisher to induce the publisher to develop additional first-copy products. When entrance to publishing is easy, numerous competing publishers make first-copy products abundant, so a long legal monopoly may do more harm by imposing deadweight loss than good by inducing more first-copy products.

The Effect of Decreasing Copying Cost b

With digital technology, the cost of reproducing and distributing copies of information products, reflected by Parameter b, is typically a few key strokes, the CPU cycles for copying, and the cost of telecommunication. Although this cost is not an absolute zero, it is much less than that of reproducing a paper book and physically transporting it to the reader. In Equation (16), a decrease in b will increase the maximum possible consumer surplus in the numerator and increase the monopoly profit and deadweight loss in the denominator. To simplify notation, let

$$W \equiv \int_{b}^{\infty} D(p)\, dp$$

and

$$w \equiv \left(1 - \frac{1-\alpha}{n}\right)\left(p^{*} - b\right)D\left(p^{*}\right) + \rho\left(\int_{b}^{p^{*}} D(p)\, dp - \left(p^{*} - b\right)D\left(p^{*}\right)\right)$$

Then one has

$$\frac{\partial}{\partial b}G(T) = \alpha G(\infty)\frac{W}{w}\left(\frac{1}{W}\frac{\partial W}{\partial b} - \frac{1}{w}\frac{\partial w}{\partial b}\right)$$

From Equation (21), the direction of change of optimal duration of copyright protection depends on the percentage change of the maximum possible consumer surplus and the percentage change of the weighted sum of the monopoly revenue, net copying and distribution cost, and deadweight loss with change in b. If these two percentage changes are in the same direction and identical in magnitude, the net effect of change in copying cost may be nil.

This is in fact the case for $g(t)$ and $D(p)$, and for any combination of $D(p)$ and $g(t)$ from $D(p) = e^{-\lambda p}$, $D(p) = p^{-\lambda}$ and $g(t) = 1 - \frac{t}{\beta}$, $g(t) = 1$ for $0 \leq t \leq \beta$, or $g(t) = e^{-\beta t}$. Thus, it seems replication and distribution costs do not directly affect optimal copyright duration. The general effect of b on optimal duration needs further investigation. However, it seems plausible that change in copying and distribution costs have little effect on optimal copyright duration if the two percentage changes in Equation (21) cancel each other.

Effect of Globalization

Networking technology leads to the ultimate globalization of information markets. This globalization has two effects. First, publishers in different countries compete more and more directly, thus the number of competing publishers is in a sense increased. Second, the level of demand for a given set of information products may increase as new markets develop overseas (i.e., D_0 is increased). It is natural to ask whether this globalization of information markets has any implication for copyright duration.

From a domestic point of view, when markets develop overseas, it is always good for the publishers of this country for foreign countries to enforce longer copyright protection. It is in the interest of the primary information exporting countries, such as the United States, to ask all foreign countries to enforce a longer copyright protection. To do so, the main information exporting countries would first legislate a longer duration of protection at home, then ask foreign countries to do the same. This longer duration should balance the interest generated by an expected similarly longer protection in foreign countries, and the consumer welfare foregone at home due to a longer copyright protection at home.

However, if all countries were equal, and each country tries to maximize its interest in enforcing copyright protection, from a pure economic point of view there are three possible equilibria: (a) no copyright is enforced in any country; (b) each country enforces copyright protection for information products developed at home, but no protection is enforced for products developed abroad; and (c) a uniform copyright protection is enforced in all countries for all products developed at home and abroad alike, which maximize the welfare of all countries as a whole. Clearly, the third is Pareto superior, and the first is the most inferior outcome for "global social welfare." Moreover, it is possible for the third to be stable in a repeated game. Thus, it seems that a "global copyright protection" may be theoretically preferred and possible.

As mentioned earlier, in a global information market brought by new information technology, the number of competing publishers is increased and the level of demand for a given set of information products is raised. Thus, the parameters n and D_0 are increased in this model. According to the results reported earlier, to the extent that the number of competing publishers is increased, the duration of copyright protection optimal from a global point of view will be decreased.

In this model, the parameter D_0 cancels out and does not appear in the result of Equation (16). Thus, according to this model, there is no need for change in the duration of copyright protection to the extent that the demand for a given set of information products is increased because of globalization of information market.

Effects of Other Factors

This section further examines how other parameters in the model might affect copyright duration.

DIFFERENTIATION OF PRODUCTS. As mentioned earlier, parameter α reflects the differentiation of information products, with $\alpha = 0$ representing identical products and perfect substitution, and $\alpha = 1$ representing no substitution. From Equation (16), optimal duration of copyright protection increases with the level of differentiation α.

For example, setting the number of competing publishers at $n = 10$, and using the earlier choices for $D(p)$, $g(t)$, and the other parameters, one can solve for optimal duration T for different values of α. The results are shown in Figure 14.2.

Thus, the more differentiated the products are, the longer the protection should be. When a more differentiated product is added to the existing pool of products, it should be more valuable, and thus deserving of more incentive for its development, implying a longer protection. This is also consistent with the effect of competition. The more differentiated the new product, the less it constitutes competition to existing products, thus the longer protection.

THE ECONOMIC LIFE OF PRODUCTS. Recall that $g(0)=1$ and $g(t)$ is a non-negative function. $G(\infty)$ is the integral of $g(t)$ from Time 0 to infinity, discounted by rate γ. The longer the economic life of the information product, and the slower the value of the product falls off over time, the bigger is $G(\infty)$. From (16), T increases with $G(\infty)$. Thus, the longer the economic life of information products, the longer the optimal copyright protection. If all information were of the "daily news" kind, copyright protection would be shorter than that deserved by Shakespeare.

With the choices of functional forms and parameter values (and $n = 10$), Figure 14.3 shows the dependence of optimal copyright duration on the economic life of information products (as measured by β). This figure shows that optimal duration generally increases with economic life β.

Figure 14.2
Effect of product differentiation on optimal duration

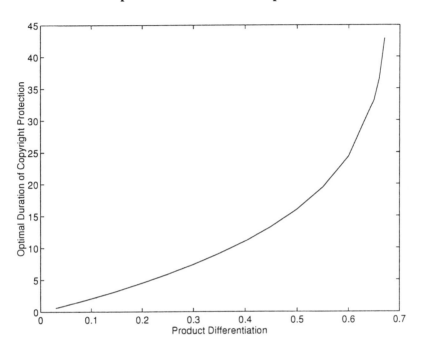

The x-axis in Figure 14.3 has been intentionally extended to show that, as the economic life of information products extends to infinity, the optimal duration of protection may not follow suit. In fact, with the parameters chosen, optimal duration increases quickly, then asymptotically approaches 28 years as β increases to infinity. This result is quite general, and is related to the discounting factor used when evaluating copies of products over time.

RETURN TO SCALE IN FIRST-COPY DEVELOPMENT. Parameter ρ represents the economy of scale in first-copy development, with $\rho > 1$ representing decreasing return and $\rho < 1$ increasing return to scale. One can expect returns to scale to decrease after some number of first-copy products (i.e., the marginal cost of first-copy development will finally increase with the number of first-copy products developed). For example, it becomes more difficult to write additional articles with the same quality after some threshold number of chapters has been written. A big-

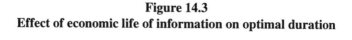

Figure 14.3
Effect of economic life of information on optimal duration

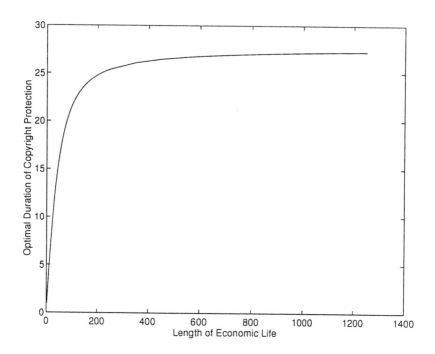

ger ρ represents that the marginal cost of first-copy development increases faster with the number of first-copy products than if ρ is smaller.

From Equation (16), optimal duration of copyright protection decreases with ρ. Figure 14.4 shows that optimal duration generally decreases when the economies of scale in first-copy development decrease. Why this is the case is not very intuitive. It seems that when the marginal cost of first-copy products increases faster, it is not so worthwhile to forego the same level of consumer welfare obtainable from the other products developed "before the margin" to induce publishers to develop the costly marginal products.

Note the return to scale, represented by ρ, is different from the general level of development cost, reflected by c. Parameter c does not appear in the formulation of optimal duration in Equation (16), indicating that it has no effect on optimal duration. The reason for the irrelevance of the general level of first-copy cost may be

Figure 14.4
Effect of economy of scale in first-copy development

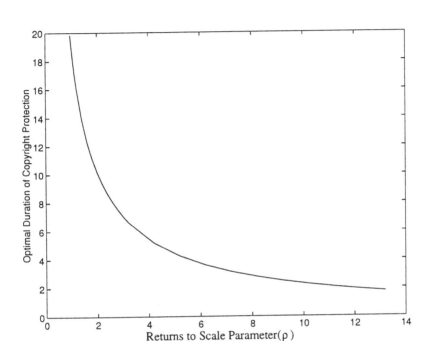

that, although a higher cost requires a longer duration of protection to induce the same level of development, the higher level of development cost means that the information is less desirable from the social welfare point of view. Thus, such products deserve a low level of development, and so their duration of protection may remain the same.

In reality, the general level of development cost is often related to the quality of the product, and thus to consumer demand for the product. If the development cost affects the shape of the demand function (e.g., changing parameter λ), then change of the level of development cost may affect the optimal duration of copyright protection. In this model, the interaction among development cost, product quality, and the shape of demand curve, and thus the indirect effect of development cost on optimal duration, is not considered. However, if the development cost, through product quality, only affects the general level of demand, not its shape, the indirect effect is nil because the effect of the general level of demand D_0 is nil.

WELFARE LOSS OF DEVIATION FROM OPTIMAL DURATION

The model presented earlier determines the optimal duration of copyright protection given the functional forms and parameter values governing information demand, and costs of development, reproduction, and distribution. The duration is optimal because it maximizes social welfare. One could ask to what extent social welfare will be sacrificed by suboptimal duration of copyright protection. The same question can be asked about the availability of first-copy information. Figure 14.5 shows the welfare loss due to suboptimal protection; Figure 14.6 shows its effect on availability of first-copy information products.

In the example, the optimal duration is about 16 years. Figure 14.5 shows that if actual duration is 10 years shorter than optimal, social welfare incurs a 16% loss; when it is 10 years longer, there is a 3% loss. It appears that erring on the side of a

Figure 14.5
Percentage loss in welfare

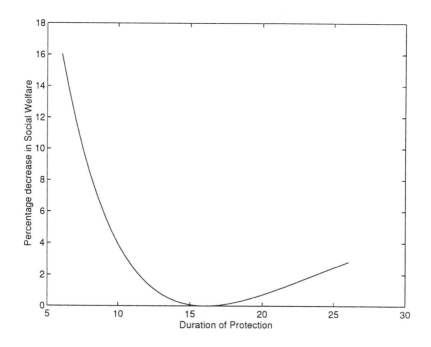

Figure 14.6
Copyright protection and information availability

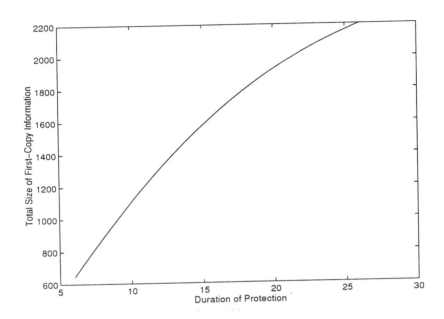

longer protection term is less injurious than missing in the other direction. This result is true in general if demand for information products decreases over time.

Figure 14.6 shows that the production of first-copy information will increase monotonically with duration of protection. Figures 14.5 and 14.6 together imply that, although availability of first-copy information will always increase with longer protection, an excessively long protection may be harmful from the social welfare point of view. Although new first-copy products induced by longer protection add profit to the publisher and surplus to consumer, the longer protection may cause more deadweight loss on other existing products.

FINAL REMARKS

This chapter analyzed the effects on optimal copyright duration of major changes in the information market caused by new information technology. First, the analysis indicates that optimal protection should be shorter because natural barriers are removed by technology development, allowing more players to enter and compete in the information marketplace. Second, from a global point of view, globalization does not affect optimal duration of copyright protection to the extent that it increases demand for a given set of information products. However, globalization calls for a shorter duration of protection to the extent that competition is enhanced internationally. Third, the reduction of reproduction and distribution costs may have little direct effect on the optimal duration of copyright, besides the indirect effect of lowering barriers to entrance, which facilitates competition and shortens optimal duration.

Thus, the model yields no support for extending duration of copyright protection to adapt to changes caused by new information technology. The main result seems to point to the other direction (i.e., to a shorter duration of copyright protection), because an increase in the number of publishers, made possible by new technology, results in greater abundance of first-copy information.

However, as mentioned earlier, these results cannot be taken as a suggestion for or an assessment of a change of copyright duration. To do that, one needs to evaluate if the existing duration is appropriate to market conditions prevailing before the changes caused by new information technology, which has not been done. Furthermore, there is another reason for information-exporting countries to extend copyright duration. It is in the interest of information-exporting countries to induce other countries to enforce a longer copyright protection. This interest, of course, should be balanced by the accompanying foregone consumer welfare at home when considering a longer copyright protection.

This chapter also analyzed the effects of some other factors on optimal duration. First, the more differentiated the information products are, the longer should be the optimal duration. Second, the optimal duration increases generally with the economic life of information. However, an infinite length of economic life does not necessarily mean an infinite duration of protection. Third, although the general level of cost in first-copy development does not affect optimal duration, optimal duration decreases with a decrease in economies of scale in first-copy development (i.e., if the marginal cost of first-copy development increases quicker with number of first-copy products, the optimal duration of copyright should be shorter).

However, the model is inevitably simple compared with the complex nature of information markets and the copyright system. For example, the model does not include non-monetary motivation for information development, nor does it consider the cost of implementing the copyright system. Both noncommercial development and the cost of enforcing copyright protection are not neutral to new technology. New technology can make information developed with nonmonetary motivation more widely available, satisfying demand for information that otherwise may only be satisfiable through commercial means. This may reduce the demand for commercial information products. Also, digital technology makes illegal reproduction and distribution easier, thus making protection more difficult. This means that enforcing copyright protection may be more costly.

However, this analysis may shed light, at least tangentially, on how some of those factors may influence the optimal duration of copyright protection. This analysis showed that change in the general level of demand for commercial information products has no effect on the optimal duration of protection. This means reduction of demand for commercial information products, due to wider availability of noncommercial products, may not affect optimal duration of protection of commercial information. Furthermore, the cost of copyright enforcement may be treated in a way similar to deadweight loss caused by copyright protection. Thus, to the extent that the cost of the copyright system is increased, because of ease of illegal reproduction and distribution, copyright protection should be shorter.

ACKNOWLEDGMENT

We wish to thank Professor Wendy Gordon of the University of Boston Law School and Professor Peggy Radin of Stanford University Law School for their valuable comments and suggestions. All errors remain ours.

REFERENCES

Brassil, J., Low, S., Maxemchuk, N., & O' Gorman, L. (1994). *Electronic marking and identification techniques to discourage document copying.* Technical Report, AT&T Bell Laboratories.

Braunstein, Y. (1980). Economics of property rights as applied to computer software and data bases. In N. Henry (Ed.), *CONTU: The future of information technology, Vol. 4 of copyright, Congress and technology: The public record,* (pp. 1–110). The Oryx Press, Phoenix, AZ.

Guinan, R. (1963). Duration of copyright. In The Copyright Society of the United States of America (Ed.), *Studies on copyright* (Vol. 1, pp. 473–494). Fred B. Rothman & Co., Hackensack, NJ.

Advisory Council on Information Highway of Canada (1995). *Copyright and the information highway: Final report of the Subcommittee on Copyright.* http://info.ic.gc.ca/info-highway/reports/copyright/copy-e.txt

Lehman, B. (1995). *Intellectual property and the National Information Infrastructure: The report of the Working Group on Intellectual Property Rights.* Information Infrastructure Task Force.

Yuan, Y., Roehrig, S., & Sirbu, M. (1995). Service models, operational decisions and architecture of digital libraries. In F.M. Shipman, R. Furuta, and D.M. Levy (Eds.), *Proceedings of Digital Libraries'95. College Station, TX, Hypermedia Research Laboratory, pp. 197–201.*

15

Revising Copyright Law
for the Information Age

Jessica Litman
Wayne State University

Our current copyright law is based on a model devised for print media, and expanded with some difficulty to embrace a world that includes live, filmed and taped performances, broadcast media, and, most recently, digital media. That much is uncontroversial. The suitability of that model for new media is much more controversial. As one might expect, to the extent that current legal rules make some parties "haves" and others "have-nots," the haves are fans of the current model,[1] while today's have-nots suggest that some other model might be more appropriate for the future.[2] Meanwhile, copyright lawyers, who, after all, make their livings interpreting and applying this long and complex body of counterintuitive, bewildering rules, insist that the current model is very close to the platonic ideal,

---

1. *See, e.g., Intellectual Property Issues Involved in the National Information Infrastructure Initiative: Public Hearing Before the National Information Infrastructure Task Force Working Group on Intellectual Property,* Nov. 18, 1993 [hereinafter *Nov. 18, 1993 IITF Hearing*], at 14 (testimony of Steven J. Metalitz, Information Industry Association); *id.* at 99 (testimony of Fritz E. Attaway, Motion Picture Association of America); *Public Hearing at The University of California, Los Angeles Before the Information Infrastructure Task Force Working Group on Intellectual Property,* September 16, 1994 [hereinafter *Sept. 16, 1994 IITF Hearing*], at 5–7 (testimony of George Vradenburg, Fox, Inc.).

2. *See, e.g., Nov. 18, 1993 IITF Hearing, supra* note 1, at 85 (testimony of David H. Rothman); *Public Hearing at The University of Chicago Before the Information Infrastructure Task Force Working Group on Intellectual Property Rights,* September 14, 1994 [hereinafter *Sept. 14, 1994 IITF Hearing*], at 24–26 (testimony of John Rademacher, American Farm Bureau Federation).

and should under no circumstances be jettisoned in favor of some untried and un-
true replacement.[3]

The Clinton Administration's Information Infrastructure Task Force has recently waded into the quagmire with a report entitled *Intellectual Property and the National Information Infrastructure*[4] ("White Paper") that seeks to have it both ways.
The White Paper concludes that so long as the meaning of the current copyright
law, and the way that law should be read to apply to new technology, is clarified,
then the current law is "fundamentally adequate and effective."[5] The White Paper,
therefore, takes on the task of interpreting current law to resolve any ambiguities
that might arise in the context of new technology. Using the tools that all lawyers

---

3. *See, e.g.*, Jane C. Ginsburg, *Four Reasons and a Paradox: The Manifest Superiority of Copyright
Over Sui Generis Protection of Computer Software*, 94 COLUM. L. REV. 2559 (1994); *Public Hearing
at Andrew Mellon Auditorium Before the Information Infrastructure Task Force Working Group on Intellectual Property Rights*, September 22, 1994 [hereinafter *Sept. 22, 1994 IITF Hearing*], at 31 (testimony of Morton David Goldberg, Schwab, Goldberg, Price and Dannay).

4. INFORMATION INFRASTRUCTURE TASK FORCE, INTELLECTUAL PROPERTY AND THE NATIONAL INFORMA-
TION INFRASTRUCTURE: THE REPORT OF THE WORKING GROUP ON INTELLECTUAL PROPERTY RIGHTS
(1995) [hereinafter WHITE PAPER]. The White Paper is not the florid endorsement of enhanced copyright protection that its predecessor draft report was. *See* INFORMATION INFRASTRUCTURE TASK
FORCE, INTELLECTUAL PROPERTY AND THE NATIONAL INFORMATION INFRASTRUCTURE: A PRELIMINARY
DRAFT OF THE REPORT OF THE WORKING GROUP ON INTELLECTUAL PROPERTY RIGHTS (1994) [here-
inafter GREEN PAPERS]. The Green Paper draft predicated its analysis on the premise that unless copyright law were strengthened, the originators of content would refuse to make their works available to
the public. *Id.* at 6–10. Its approach was twofold. First, the draft contained revisionary interpretations
of current law that enhanced copyright owners' control over their works. *See id.* at 35–37, 45–53. Second, the report gave suggestions for further fortifying that control. *Id.* at 120–33. *See generally, e.g.*,
Jessica Litman, *The Exclusive Right to Read*, 13 CARDOZO ARTS & ENT. L.J. 29 (1994); Pamela Samuelson, *Legally Speaking: The NII Intellectual Property Report*, 37 COMMUNICATIONS OF THE ACM,
December 1994, at 21; Diane Leenheer Zimmerman, *Copyright in Cyberspace: Don't Throw Out the
Public Interest with the Bath Water*, 1994 ANN. SURV. AM. L. 403.

The White Paper spends most of its ink on the revisionary interpretation leg of the strategy: it asserts
that most of the enhanced protection copyright owners might want is *already available* under current
law, at least so long as that law is properly interpreted, and it contains a long exegesis of what the properly interpreted copyright law should be read to provide. *See* WHITE PAPER, *supra*, at 19–130. The
difference is largely one of style rather than substance, as the White Paper ends up endorsing most of
the recommendations tentatively included in the Green Paper, but its tone is noticeably less hostile to
the universe of users of copyrighted works. The White Paper and Green Paper, and transcripts of the public hearings held in September, 1994, *supra* notes 1–3, are available electronically via the Task Force's
gopher server at iitf.doc.gov [URL gopher://ntiaunix1.ntia.doc.gov:70/11s/iitf/infopol]. The November
18, 1993 Hearing transcript, *supra* note 1, was initially released on the IITF gopher and then removed;
a copy of that transcript is preserved at URL gopher://sunbird.usd.edu:70/11/Academic%20Divisions/
School%20of%20Law/NII%20Working%20Group%20on%20Intellectual%20 Property.

5. WHITE PAPER, *supra* note 4, at 212.

use when engaged in such tasks,[6] the White Paper carefully explains that just about every ambiguity one can imagine, properly understood, should under the best view of current law be resolved in favor of the copyright holder.

That leads the authors of the White Paper to some pretty startling conclusions. Most notably, since any use of a computer to view, read, reread, hear or otherwise experience a work in digital form requires reproducing that work in a computer's memory, and since the copyright statute gives the copyright holder exclusive control over reproductions,[7] everybody needs to have either a statutory privilege or the copyright holder's permission to view, read, reread, hear or otherwise experience a digital work, each time she does so.[8] Not only individuals, but their Internet service providers and the proprietors of any computers that assist in the transfer of files are, and should be, liable for copyright infringement in these cases, regardless of whether they know someone's intellectual property rights are being invaded, or even what content is moving through their equipment.[9] Once it is understood that current copyright law in fact so provides, the White Paper argues, there is little need to amend it to make express provision for new technology; only minor adjustments will be required.[10] Thus, the White Paper neatly avoids addressing the policy question whether copyright should be defined in terms that convert individ-

---

6. When convenient for its argument, the White Paper relies on the expressed intent of congressional committees to buttress its analysis of current copyright law. *See, e.g.*, WHITE PAPER, *supra* note 4, at 226. When express language in the congressional Committee Reports is less convenient, the White Paper ignores it, *see, e.g., id.* at 65, or characterizes it as irrelevant, *see id.* at 72 n.226. Similarly, the White Paper is selective in its citation of case authority. *See* James Boyle, *Overregulating the Internet*, WASH. TIMES, Nov. 14, 1995, at A17; Pamela Samuelson, *The Copyright Grab*, WIRED, January, 1996, at 137-38.

7. 17 U.S.C. § 106 (1994):
[T]he owner of a copyright . . . has the exclusive rights to do and to authorize any of the following:
(1) to reproduce the copyrighted work in copies or phonorecords.
"Copies" are defined in 17 U.S.C. § 101 as
material objects . . . in which a work is fixed by any method now known or later developed, and from which the work can be perceived, reproduced, or otherwise communicated, either directly or with the aid of a machine or device.
. . .
"Fixed" is also defined in section 101:
A work is "fixed" in a tangible medium of expression when its embodiment in a copy or a phonorecord . . . is sufficiently permanent or stable to permit it to be perceived, reproduced, or otherwise communicated for a period of more than transitory duration. . . .

8. WHITE PAPER, *supra* note 4, at 64–66. This particular piece of revisionist interpretation is irresponsible. The legislative materials accompanying the Copyright Act make it clear that Congress intended to assimilate the appearance of a work (or portions of a work) in a computer's random access memory to unfixed, evanescent images rather than "copies." *See* H.R. REP. NO. 1476, 94th Cong., 2d Sess. 52–53, 62 (1976); Litman, *supra* note 4; Samuelson, *supra* note 4.

9. WHITE PAPER, *supra* note 4, at 114–24; *see* Samuelson, *supra* note 6.

10. *See, e.g.*, WHITE PAPER, *supra* note 4, at 90–95.

ual users' reading of files into potentially infringing acts, by insisting that Congress chose to set it up this way when it enacted the current law.[11]

I am sometimes a legislative historian, and the contours of this dispute don't look very different from the shape of very similar disputes that arose in the 1980s, when the gods invented personal computers; or the 1970s, when they invented videocassette recorders; or the 1960s, when they invented cable television; or the 1920s, when they invented commercial broadcasting and talkies. Arguing that Congress already considered a question, and resolved it in one's favor then, is a common tactic in the history of copyright lobbying because it bypasses the problem of persuading Congress to consider the question and resolve it in one's favor today.[12] Sometimes it works; other times it fails. In evaluating these claims, it is always useful to inject a note of realism: would Congress have adopted such-and-such language if it believed at the time the legislation was enacted that this language would be interpreted to mean what is now being claimed? Whether a platonic Congress would have made that call or not, in view of what we now know about how the world has evolved, is that choice a good one, in policy terms? People are going to differ on the answers to both of these questions, but at least their differences are on the table; we aren't making information policy by sleight-of-hand.

Until now, our copyright law has been addressed primarily to commercial and institutional actors who participated in copyright-related businesses.[13] The statute seems on its face to have been drafted primarily for the benefit of people with ready access to copyright counsel. It is long, complicated, counterintuitive and highly specific.[14] It has very few words to say to the non-commercial, non-institutional user interacting with copyrighted works in his or her private capacity. What it does say seems to boil down to this:

> The copyright owner has some very broad exclusive rights. Those rights
> have never extended to private performances or displays, or to resale of
> copies of works, but they do include command over reproduction, adapta-
> tion, distribution, public performance or display. All of those rights are
> subject to a host of explicit exceptions, but none of those exceptions ap-
> plies to you.

---

11. WHITE PAPER, *supra* note 4, at 64–66, 90–95, 214–17. *See* Leslie A. Kurtz, *Copyright and the National Information Infrastructure in the United States*, 18 EUR. INTELL. PROP. REV. 120 (1996).

12. *See* Jessica Litman, *Copyright Legislation and Technological Change*, 68 OR. L. REV. 275, 353–54 (1989); *see, e.g., Home Video Recording: Hearings Before the Senate Judiciary Comm.*, 99th Cong., 2d Sess. 2, 7–8 (1987) [hereinafter *Home Video Recording*] (testimony of Jack Valenti, President/Chief Executive Officer, Motion Picture Association of America); *id.* at 84–94 (statement of Charles D. Ferris, Home Recording Rights Coalition); *see generally Home Audio Recording Act: Hearings on S. 1739 before the Senate Comm. on the Judiciary*, 99th Cong., 1st & 2d Sess. (1986).

13. *See* PAUL GOLDSTEIN, COPYRIGHT'S HIGHWAY 129–64 (1994).

14. *See* Litman, *supra* note 4.

How can this be? It results from the process we rely on to write our copyright laws. The current federal copyright statute (and its predecessors) were composed by representatives of copyright-related industries to govern interactions among them. We have built into the process a mechanism for the cable television industry, or the software publishers' association, or the manufacturers of digital audio tape to insist that the law include a provision privileging this or that use that that party deems essential. We have never had a mechanism for members of the general public to exert influence on the drafting process to ensure that the statute does not unduly burden private, non-commercial, consumptive use of copyrighted works. The design of the drafting process (in which players with major economic stakes in the copyright sphere are typically invited to sit down and work out their differences *before* involving members of Congress in any new legislation) excludes ordinary citizens from the negotiating table.[15]

For that reason, it should be unsurprising that many members of the public believe that copyright law simply doesn't apply to ordinary folks making non-commercial use of copyrighted works.[16] It should also be unsurprising that many copyright holders insist that current copyright law applies to everyone as literally written; there are no privileges or exemptions for the public because it would be inappropriate for members of the public to have them.[17]

Most of the current proposals for interpreting or amending copyright law to apply to digital technology have come from copyright holders or their defenders. They appear to contemplate that the law will henceforth govern all activities of anyone who uses digital media in any capacity. The suggested application of copyright law to digital media potentially affects, indeed prohibits, most common everyday ways one can use a computer or other digital device to read, view, hear or otherwise experience copyrighted works. Under one controversial view of the copyright statute, most of the activity that takes place over the Internet, and much of the activity that takes place in individuals' homes when they turn on their computers, cannot lawfully be engaged in without the authorization of the copyright owner in the material they see, hear, read, listen to or view. It is that view of the

---

15. *See id.* at 50–53. *See generally* 1 COPYRIGHT, CONGRESS AND TECHNOLOGY: THE PUBLIC RECORD xi–xxii, 1–5, 99–105 (Nicholas Henry ed. 1978); 2 COPYRIGHT, CONGRESS AND TECHNOLOGY, *supra*, at xv–xxiii (1979); 3 COPYRIGHT, CONGRESS AND TECHNOLOGY, *supra*, at xiii–xli (1980); Litman, *supra* note 12; *see also* Thomas P. Olson, *The Iron Law of Consensus: Congressional Responses to Proposed Copyright Reforms Since the 1909 Act*, 36 J. COPYRIGHT SOC'Y 109 (1989).

16. *See* OFFICE OF TECHNOLOGY ASSESSMENT, U.S. CONGRESS, COPYRIGHT & HOME COPYING: TECHNOLOGY CHALLENGES THE LAW 163–65 (1989); OFFICE OF TECHNOLOGY Assessment, U.S. CONGRESS, INTELLECTUAL PROPERTY RIGHTS IN AN AGE OF ELECTRONICS AND INFORMATION 121–23, 208–09 (1986); Pamela Samuelson, *Fair Use for Computer Programs and Other Copyrightable Works in Digital Form: The Implications of Sony, Galoob, and Sega*, 1 J. INTELL. PROP. L. 49, 67 (1993).

17. *See, e.g.*, WHITE PAPER, *supra* note 4 , at 73 n.227; *Sept. 14, 1994 IITF Hearing, supra* note 2, at 44–46 (testimony of Edward Massie, Commerce Clearing House); June M. Besek, *Future Copyright Protection: Is Existing Law Adequate in a Networked World?*, N.Y.L.J., Dec. 5, 1994, at Supp. 1.

law that the Clinton Administration's Information Infrastructure Task Force has gotten behind.[18]

I have argued elsewhere that a lot of what is in the Working Group's Report is disingenuous and much of the rest of it is simply wrong.[19] I won't rehearse those arguments here. I do think, however, that the Working Group can be accused of not taking sufficiently seriously its assignment to consider whether the copyright law *should* be revised, and if so, how;[20] it got caught up in the politics of its situation, and spent its energies trying to ensure that all current stakeholders and potential campaign contributors would be better off under the administration's proposals than they were under the status quo. I propose to give the question more serious consideration.

I

When we examine the question whether copyright needs redesign to stretch it around digital technology, we can look at the issues from a number of different vantage points. First, there is the viewpoint of current copyright stakeholders: today's market leaders in copyright-affected industries. Their businesses are grounded on current copyright practice; their income streams rely on current copyright rules. Most of them would prefer that the new copyright rules for new copyright-affecting technologies be designed to enable current stakeholders to retain their dominance in the marketplace.[21]

One way to do that is to make the new rules as much like the old rules as possible. Current copyright holders and the industries they do business with are already set up to operate under those rules: they have form agreements and licensing agencies and customary royalties in place. There are other advantages in using old rules: if we

---

18. *See* WHITE PAPER, *supra* note 4, at 63–96.

19. *See* Litman, *supra* note 4; *Sept. 22, 1994 IITF Hearing, supra* note 3, at 66–67 (testimony of Jessica Litman); Jessica D. Litman, 13 THE WAYNE LAWYER 18 (Summer 1995) (excerpt from her speech, The National Information Infrastructure: How Computers Are Changing Intellectual Property Law, Address Before the Association of American Law Schools Section on Computer Law (Jan. 8, 1995)).

20. *See* Kurtz, *supra* note 11, at 121–23, 126.

21. Note, here, that we are talking not only about author-stakeholders, or publisher-stakeholders, but also about collecting-agency-stakeholders. The dispute between the Harry Fox Agency and ASCAP over whether a transmission over the Internet should be deemed a distribution or a performance is exemplary: The composer would get the royalties either way, but the collecting entity's cut would go to a different stakeholder. *See Public Hearing at Andrew Mellon Auditorium Before the Information Infrastructure Task Force Working Group on Intellectual Property Rights,* September 23, 1994 [hereinafter *Sept. 23, 1994 IITF Hearing*], at 19–22 (testimony of Stu Gardner, composer); *id.* at 25–28 (testimony of Michael Pollack, Sony Music Entertainment); *id.* at 28–31 (testimony of Marilyn Bergman, ASCAP); *id.* at 31–33 (testimony of Hillary Rosen, Recording Industry Association of America); *id.* at 33–38 (testimony of Frances Preston, BMI); *id.* at 38–43 (testimony of Edward Murphy, National Music Publishers' Association).

treat the hypertext version of the New York Times as if it were a print newspaper, then we have about 200 years worth of rules to tell us how to handle it. We can avoid the problems that accompany writing new rules, or teaching them to the people (copyright lawyers, judges, newspaper publishers) who need to learn them.

Using old rules, however, has the obvious disadvantage that they will not necessarily fit the current situation very well. Where the new sorts of works behave differently from the old sorts of works, we need to figure out some sort of fix. Here's a simple example: Newsstands turn out to be an effective way of marketing literary works in part because it is difficult as a practical matter to make and distribute additional copies of newspapers and magazines that one buys from the newsstand. If one "buys" a newspaper by downloading it from the World Wide Web, on the other hand, it is pretty easy to make as many copies as one wants. The old rules, customs and practices, therefore, will not work very well unless we can come up with a way to prevent most of those copies from getting made. Relying on old rules encourages us to solve the problem that the World Wide Web is not like a newsstand by disabling some of its un-newsstand-like qualities. We could enact rules requiring the proprietors of web pages to set them up to behave much more like newsstands; we could demand that they insert code in each of their documents that would prevent downloading or would degrade any downloaded copies; we could require modem manufacturers to install chips that disabled the transfer of digital data unless some credit card were charged first.[22]

But would we really want to do that? Adopting rules that disable new technology is unlikely to work in the long term, and unlikely to be a good policy choice if it does work. We have tried before to enact laws that erect barriers to emerging technology in order, for policy reasons, to protect extant technology. The FCC did precisely that when it regulated cable television to the point of strangulation in order to preserve free broadcast TV. That particular exercise didn't work for very long.[23] Others have been more successful. Direct broadcast satellite television is still not widely available in the U.S., and no small part of the reason is that our current legal infrastructure makes it much more difficult for direct satellite broadcasters than for cable operators or conventional broadcasters.[24]

If our goal in reforming current law were to make things more difficult for emerging technology in order to protect current market leaders against potential competition from purveyors of new media, then cleaving to old rules would be a satisfactory, if temporary, solution. Adhering to old rules might distort the marketplace for new technology for at least the short term (since that, after all, would be one of its purposes), which might influence how that technology developed in the

---

22. *Cf.* WHITE PAPER, *supra* note 4, at 230–34 (recommending prohibition of devices and services that defeat copy protection).

23. *See* Jonathan Weinberg, *Broadcasting and the Administrative Process in Japan and the United States*, 39 BUFF. L. REV. 615, 694–700 (1991).

24. *See* Litman, *supra* note 12, at 342–46.

longer term, which, in turn, might influence whether and how the affected industries would compete in the markets for those technologies in the future. It would probably delay the moment at which the current generation of dominant players in information and entertainment markets were succeeded by a new generation of dominant players in different information and entertainment markets.

II

If, instead of looking at the situation from the vantage point of current market leaders, we imagined the viewpoint of a hypothetical benevolent despot with the goal of promoting exciting new technology, we might reach an entirely different answer to the question. Such a being might look at history and recognize that copyright shelters and exemptions have, historically, encouraged rapid investment and growth in new media of expression. Player pianos took a large bite out of the markets for conventional pianos and sheet music after courts ruled that making and selling piano rolls infringed no copyrights;[25] phonograph records supplanted both piano rolls and sheet music with the aid of the compulsory license for mechanical reproductions;[26] the jukebox industry was created to exploit the copyright exemption accorded to the "reproduction or rendition of a musical composition by or upon coin-operated machines."[27] Radio broadcasting invaded everyone's living rooms before it was clear whether unauthorized broadcasts were copyright infringement;[28] television took over our lives while it still seemed unlikely that most television programs could be protected by copyright.[29] Videotape rental stores sprang up across the country shielded from copyright liability by the first sale doctrine.[30] Cable television gained its initial foothold with the aid of a copyright exemption[31]

---

25. *See* Kennedy v. McTammany, 33 F. 584 (C.C.D. Mass. 1888); White-Smith Music Publishing Co. v. Apollo Co., 209 U.S. 1 (1908).

26. Copyright Act of March 4, 1909, ch. 320, §§ 1(e), 35 Stat. 1075 (1909) (codified as amended at 17 U.S.C. § 1(e), 101(e)), repealed by Pub. L. No. 94-553, 90 Stat. 2541 (1976). *See* U.S. LIBRARY OF CONGRESS, SECOND SUPPLEMENTARY REPORT OF THE REGISTER OF COPYRIGHTS ON THE GENERAL REVISION OF THE U.S. COPYRIGHT LAW: 1975 REVISION BILL at Ch. IX (1975).

27. Copyright Act of March 4, 1909, ch. 320, § 1(e), 35 Stat. 1075 (1909). *See* U.S. LIBRARY OF CONGRESS, SECOND SUPPLEMENTARY REPORT OF THE REGISTER OF COPYRIGHTS, *supra* note 26, at Ch. X.

28. *See* Litman, *supra* note 12, at 291–304.

29. *See id.* at 305–08.

30. *See* 17 U.S.C. § 109 (1994). *See generally Home Video Recording*, *supra* note 12, at 2–3.

31. *See* Teleprompter Corp. v. Columbia Broadcasting Sys., Inc., 415 U.S. 394 (1974); Fortnightly Corp. v. United Artists Television, Inc., 392 U.S. 390 (1968).

and displaced broadcast television while sheltered by the cable compulsory license.[32]

Why would a copyright exemption promote development? Conventional wisdom tells us that, without the incentives provided by copyright, entrepreneurs will refuse to invest in new media. History tells us that they do invest without paying attention to conventional wisdom. A variety of new media flourished and became remunerative when people invested in producing and distributing them first, and sorted out how they were going to protect their intellectual property rights only after they had found their markets. Apparently, many entrepreneurs conclude that if something is valuable a way will be found to charge for it, so they concentrate on getting market share first, and worry about profits—and the rules for making them—

---

32. *See* 17 U.S.C. § 111(d) (1994). Assumptions of copyright immunity can stimulate nascent businesses even if they're wrong. A widely held view about the scope of educational fair use had a great deal to do with the success of the commercial copy-shop industry. *See* Basic Books, Inc. v. Kinko's Graphics Corp., 758 F. Supp. 1522 (S.D.N.Y. 1991); Princeton University Press v. Michigan Document Servs., Inc., 855 F. Supp. 905 (E.D. Mich. 1994), *rev'd*, 74 F.3d 1512 (6th Cir. 1996), *vacated en banc*, 74 F.3d 1528 (6th Cir. 1996). The copy-shops took advantage of their supposed privilege to develop a market that publishers were ignoring; when the course-pack services proved popular, publishers found it worthwhile to set up course-pack photocopy permissions bureaus. *See* Basic Books, 758 F. Supp. 1522; American Geophysical Union v. Texaco, Inc., 60 F.3d 913 (2d Cir. 1994). Courts have split on whether that fair use claim is legitimate. *Compare Kinko's*, 758 F. Supp. at 1530–35, *with Michigan Document Servs.*, 74 F.3d at 1518–24.

Another example can be seen in the explosive growth of both commercial and non-commercial Internet services. Many subscribers assume that material that is made freely available over the Internet is not subject to any copyright restrictions except those explicitly reserved in the text of the electronic documents. Anecdotal evidence suggests that unauthorized copying and dissemination of protected material is widespread. Copyright holders testifying before the Working Group on Intellectual Property suggested that without effective copyright protection, copyright holders would refuse to make material available over the Internet, and that the information highway would have no travelers on it because there would be nothing there to read, hear or see. *See, e.g., Nov. 18, 1993 IITF Hearing, supra* note 1, at 14 (testimony of Steven J. Metalitz, Information Industry Association); *Sept. 14, 1994 IITF Hearing, supra* note 2, at 20 (testimony of Priscilla Walter, Gardner, Carton & Douglas, Chicago); *id.* at 31–32 (testimony of James Schatz, West Publishing Co.); *see also* WHITE PAPER, *supra* note 4, at 10–11. Yet today, most of the major copyright stakeholders whose lawyers made those arguments are not waiting to see whether we will succeed in shoring up our copyright protection; they have already established a presence on the Internet. Some of them have World Wide Web pages; others have made their products available through a commercial service provider like America Online; still others are making their wares generally available at no charge to anyone who has Internet access. *See* Denise Caruso, *Digital Commerce: A Study by A.C. Nielsen Seeks to Separate Buyers from Browsers on the Internet*, N.Y. TIMES, Aug. 21, 1995, at D3.

later.[33] In addition, by freeing content providers from well-established rules and customary practices, a copyright shelter allows new players to enter the game.[34] The new players have no vested interest–yet–so they are willing to take more risks in the hope of procuring one. They end up exploring different ways of charging for value. Radio and television broadcast signals are given to their recipients for free; broadcasters have figured out that they can collect money based on the size and demographics of their audiences. Many valuable software programs obtained their awesome market share by being passed on to consumers at no extra cost (like Windows 3.1® and DOS®) or deliberately given away as freeware (like Netscape®[35] or Eudora®). Other software programs may well have achieved their dominant market position in part by being illicitly copied by unlicensed users (WordPerfect® comes to mind). Indeed, industry observers agree that at least half of all of the copies of software out there are unauthorized, yet the software market is booming; it is the pride of the U.S. Commerce Department. Perhaps all of the unauthorized copies are part of the reason why.

Our hypothetical benevolent despot, then, might propose a temporary period during which the Internet could be a copyright-free zone. Nobody seems to be

---

33. Consider one example: when the IITF Working Group on Intellectual Property held its first public hearing, representatives of a variety of copyright-intensive industries argued that without stronger copyright protection, content providers would refuse to make works available in electronic form. *See Nov. 18, 1993 IITF Hearing, supra* note 1. The Green Paper used these concerns as the predicate for its suggested reforms. *See* GREEN PAPER, *supra* note 4, at 6–7. In the two years between the initial Hearing and the release of the Working Group's Final Report, the World Wide Web exploded: a variety of information and entertainment companies scrambled to establish conspicuous presences on the Web without waiting for copyright reform. Some of these companies doubtless restrict the material they make available because of fears of unbridled copying. Others, however, seem to be more interested in gaining market share than in preserving the inviolability of their content. In apparent response, the White Paper discounted the claim made in the Green Paper that, absent strong copyright protection, "the potential of the NII will not be realized," *see* GREEN PAPER, *supra* note 4, at 6, to the more moderate assertion that "the *full* potential of the NII will not be realized." WHITE PAPER, *supra* note 4, at 10 (emphasis added).

The sort of marketplace that grows up in the shelter of a copyright exemption can be vibrant, competitive, and sometimes brutal. Some prospectors will seek to develop market share on a hunch; others from conviction. Still others may aim only to generate a modestly valuable asset that will inspire some bigger fish out there to eat them. In any event, new products may be imagined, created, tested and introduced, and new media may be explored. Fierce competition is not very comfortable, but it can promote the progress of science nonetheless.

34. Indeed, one could argue that one of the virtues of established rules and practices for those that subscribe to them is the entry barriers they represent for outsiders.

35. Netscape managed to amass a 75% market share by making its software available to users for free. Although it had not yet earned a profit, a public offering of Netscape stock in August, 1995 set Wall Street records. *See* Laurence Zuckerman, *With Internet Cachet, Not Profit, A New Stock is Wall Street's Darling*, N.Y. TIMES, Aug. 10, 1995, at A1, D5.

making that sort of proposal these days,[36] so perhaps I am mistaken about what a wise ruler would view as good policy. Or perhaps all the benevolent despots in the neighborhood are off duty, on vacation, or just simply hiding. Perhaps they've sought alternative employment. Where might one find a benevolent despot if one wanted to ask questions about wise copyright policy? Conventionally, one might expect some government entity to take that approach. The generalist federal government entities, though–Congress and the White House–have historically been neither expert nor interested in copyright. The specialist agencies with copyright expertise include the Register of Copyrights in the Library of Congress and, new with the current administration, the Patent and Trademark Office. The Register of Copyrights has a tradition of extreme caution in taking policy roles. For most of its history, the Copyright Office has played the role of go-between, translator and emissary between copyright industries and Congress.[37] The Patent and Trademark Office, historically uninterested in copyright, has recently taken on an active policy role.[38] The Commissioner's work product on these issues,[39] however, suggests that his office is entirely in thrall to a constituency of current stakeholders.[40]

---

36. The White Paper speaks dismissively of "[s]ome" who "assert that copyright protection should be reduced in the NII environment." WHITE PAPER, *supra* note 4, at 14. But few such persons appear to have testified at any of the five public hearings the Working Group held. One was Ellen Kirsch, testifying for America Online that "[t]he on-line computer service industry is seriously concerned that the copyright laws currently in effect will stifle our industry and this important new method of communication and content delivery." *Sept. 22, 1994 IITF Hearing, supra* note 3, at 25. There were also a very few witnesses who suggested that the Working Group's interpretation of current law, and its suggestions for further amendments, represented unwarranted expansions of copyright protection. *See, e.g., id.* at 59–61 (testimony of Robert Oakley, American Association of Law Libraries); *Sept. 14, 1994 IITF Hearing, supra* note 2, at 5–9 (testimony of Edward Valauskas, American Library Association).

37. *See, e.g.,* Thorvald Solberg, *Copyright Law Reform,* 35 YALE L.J. 48 (1925).

38. *See* Mark Walsh, *Defending Her Turf: As Copyright Chief Marybeth Peters Rebuilds Her Staff, She Must Also Fight for Power with Patent Czar Bruce Lehman,* THE RECORDER, April 19, 1995, at 1.

39. *See* GREEN PAPER, *supra* note 4; WHITE PAPER, *supra* note 4; *see, e.g., Sept. 22, 1994 IITF Hearing, supra* note 3, at 32–33 (remarks of Commissioner Lehman); *Sept. 23, 1994 IITF Hearing, supra* note 21, at 11, 24–25 (remarks of Commissioner Lehman).

40. Appointed to chair a Working Group on Intellectual Property and the NII, Commissioner Lehman solicited industry views and requests, *see Nov. 18, 1993 IITF Hearing, supra* note 1, and issued a Preliminary Report, *see* GREEN PAPER, *supra* note 4, that endorsed everything on the industry wish list and included a widely criticized description of the many ways in which current law established much of that wish list already. The recently released Final Report, *see* WHITE PAPER, *supra* note 4, is more carefully written and reasoned, but makes essentially the same recommendations as the earlier draft.

Without a Government entity to take the benevolent despot's role, that viewpoint is being unevenly pressed by academics, policy bodies, and current stakeholders who find themselves disadvantaged or threatened by the present regime. It is not my purpose to claim that role, or to justify my own proposals on that basis.

III

A number of alternative viewpoints are possible: I'd like to focus on a third. Copyright is said to be a bargain between the public and copyright holders.[41] So far, I, and the stakeholders and government bodies I have described, have focused on the copyright holder's side of that bargain. Copyright owners, however, have never been entitled to control all uses of their works. Instead, Congress has accorded copyright owners some exclusive rights, and reserved other rights to the general public.[42] Commonly, copyright theorists assess the copyright bargain by asking whether it provides sufficient incentives to prospective copyright owners.[43] Indeed, it is conventional to argue that copyright holders should receive only such incentives as are necessary to impel them to create and disseminate new works.[44] That analysis is less than helpful, though, as appears when one tries to quantify the degree to which incentives are required. The questions "How many people who do not currently compose symphonic music would do so if symphonic music paid better?" and "How many current composers would write more stuff if there were more money in it?" turn out to be imponderable and untestable. Common experience tells us that the unpredictability of proceeds that will flow from any particular copyrighted work make a quantitative measure of the copyright incentive a poor gauge of what authors will create, or even what investors will invest in the exploitation of works protected by copyright. Economists tell us nonetheless that, at the margin, there is always an author who will be persuaded by a slight additional incentive to create another work, or who will be deterred from creating a particular work by a diminution in the copyright bundle of rights. If we rely on the simple economic model, we are led to the conclusion that every enhancement of the rights in the copyright bundle is necessary to encourage the creation of *some* work of authorship.

Asking, "What should copyright holders receive from this bargain? What do they need? What do they deserve"?, then, may be less than helpful. We might instead look at the other side of the equation and ask "What is it the public should get from the copyright bargain? What does the public need, want, or deserve?" The

---

41. *See, e.g.,* Sony Corp. of America v. Universal City Studios, Inc., 464 U.S. 417, 429 & n.10 (1984) (quoting H.R. REP. NO. 2222, 60th Cong., 2d Sess. (1909)); Kurtz, *supra* note 11, at 121; *see generally* Litman, *supra* note 4.

42. *See generally* L. RAY PATTERSON & STANLEY W. LINDBERG, THE NATURE OF COPYRIGHT: A LAW OF USERS' RIGHTS (1991).

43. *See, e.g.,* Rochelle Cooper Dreyfuss, *The Creative Employee and the Copyright Act of 1976*, 54 U. CHI. L. REV. 590 (1987); Jane C. Ginsburg, *Creation and Commercial Value: Copyright Protection of Works of Information*, 90 COLUM. L. REV. 1865, 1907–16 (1990); Wendy J. Gordon, *Fair Use as Market Failure: A Structural and Economic Analysis of the Betamax Case and its Predecessors*, 82 COLUM. L. REV. 1600 (1982); Linda J. Lacey, *Of Bread and Roses and Copyrights*, 1989 DUKE L.J. 1532 (1989).

44. *See, e.g.,* William W. Fisher III, *Reconstructing the Fair Use Doctrine*, 101 HARV. L. REV. 1661 (1988).

public should expect the creation of more works, of course, but what is it that we want the public to be able to do with those works?

The constitutional language from which Congress's copyright enactments flow describes copyright's purpose as "[t]o promote the Progress of Science and useful Arts."[45] We can begin with the assertion that the public is entitled to expect *access* to the works that copyright inspires. That assertion turns out to be controversial. Public access is surely not necessary to the progress of science. Scientists can build on each others' achievements in relative secrecy. Literature may flourish when authors have the words of other authors to fertilize their own imaginations, but literature may thrive as well when each author needs to devise her own way of wording. If we measure the progress of science by the profits of scientists, secrecy may greatly enhance the achievements we find.

Still, if valuable works of authorship were optimally to be kept secret, there would be no need for incentives in the copyright mold of exclusive rights. Authors could rely on self-help to maintain exclusive control of their works. Copyright makes sense as an incentive if its purpose is to encourage the dissemination of works in order to promote public access to them. It trades a property-like set of rights precisely to encourage the holders of protectable works to forgo access restrictions in aid of self-help. For much of this country's history, public dissemination was, except in very limited circumstances, a condition of copyright protection.[46] While no longer a condition, it is still fair to describe it as a goal of copyright protection.[47]

But why is it that we want to encourage dissemination? What is it we want the public to be able to do with these works that we are bribing authors to create and make publicly available? We want the public to be able to read them, view them and listen to them. We want members of the public to be able to learn from them: to extract facts and ideas from them, to make them their own, and to be able to build on them. That answer leads us to this question: how can we define the compensable units in which we reckon copyright protection to provide incentives (and, since the question of how much incentive turns out to be circular, let's not worry about that for now) for creation and dissemination, while preserving the public's opportunities to read, view, listen to, learn from and build on copyrighted works?

In 1790, Congress struck this balance by limiting the compensable events within the copyright owner's bundle of rights to printing, reprinting, publishing and vend-

---

45. U.S. CONST. art. I, § 8, cl. 8.

46. The 1976 Copyright Act extended federal statutory copyright to unpublished works. Before that, copyright protection was available for published works and for works, such as lectures or paintings, that were typically publicly exploited without being reproduced in copies. *See generally* 1 WILLIAM F. PATRY, COPYRIGHT LAW AND PRACTICE 414–21 (1994).

47. *See* L. Ray Patterson, *Copyright and "The Exclusive Right" of Authors*, 1 J. INTELL. PROP. L. 1, 37 (1993).

ing copyrighted works.[48] (That translates, in current lingo, into an exclusive right to make, distribute and sell "copies.")[49] Public performances, translations, adaptations, and displays were all beyond the copyright owner's control. Courts' constructions of the statute supplied further limitations on the copyright owner's rights. The statutory right to vend was limited by the first sale doctrine.[50] The statutory rights to print and reprint did not apply to translations and adaptations,[51] did not prevent others from using the ideas, methods or systems expressed in the protected works,[52] and, in any event, yielded to a privilege to make fair use of copyrighted works.[53]

Congress, over the years, expanded the duration and scope of copyright to encompass a wider ambit of reproduction, as well as translation and adaptation, public for-profit performance, and then public performance and display. It balanced the new rights with new privileges. Jukebox operators, for example, enjoyed an exemption from liability for public performance for more than fifty years, and were the beneficiaries of a compulsory license for another decade after that.[54] Other compulsory licenses went to record companies,[55] cable television systems,[56] satellite carriers[57] and noncommercial television.[58] Broadcasters received exemptions permitting them to make "ephemeral recordings" of material to facilitate its broadcast;[59] manufacturers of useful articles embodying copyrighted works received a flat exemption from the reproduction and distribution rights to permit them to advertise their wares.[60] Libraries received the benefit of extensive privileges to duplicate copyrighted works in particular situations.[61] Schools got an express privilege to perform copyrighted works publicly in class;[62] music stores got

---

48. Act of May 31, 1790, ch. 15, § 1, 1 Stat. 124 (1790).

49. *See* 17 U.S.C. § 106(1), (3) (1994).

50. Bobbs Merrill Co. v. Straus, 210 U.S. 339 (1908); Harrison v. Maynard, Merril & Co., 61 F. 689 (2d Cir. 1894). The first sale doctrine allows the owner of any lawful copy of a work to dispose of that copy as she pleases.

51. Stowe v. Thomas, 23 F. Cas. 201 (C.C.E.D. Pa. 1853) (No. 13,514); Kennedy v. McTammany, 33 F. 584 (C.C.D. Mass. 1888).

52. Baker v. Selden, 101 U.S. 99 (1879).

53. *See* Folsom v. Marsh, 9 F. Cas. 342 (C.C.D. Mass. 1841) (No. 4,901).

54. *See* 2 PATRY, *supra* note 46, at 971–87.

55. *See* 17 U.S.C. § 115 (1994).

56. *See id.* § 111(c).

57. *See id.* § 119.

58. *See id.* § 118.

59. *See id.* § 112.

60. *See id.* § 113(c).

61. *See id.* § 108.

62. *See id.* § 110(1).

an express privilege to perform music publicly in their stores;[63] and small restaurants got an express privilege to perform broadcasts publicly in their restaurants.[64] Congress did not incorporate specific exemptions for the general population in most of these enactments because nobody showed up to ask for them.[65] At no time, however, did Congress or the courts cede to copyright owners control over looking at, listening to, learning from or *using* copyrighted works.

IV

The right "to reproduce the copyrighted work"[66] is commonly termed the fundamental copyright right.[67] The control over the making of copies is, after all, why this species of intellectual property is called a *copy*right. So it is tempting, and easy, to view the proliferation of copying technology as threatening copyright at its core.[68] However we revise the copyright law, many argue, we need to ensure that the copyright owner's control over the making of every single copy of the work remains secure. This is especially true, the argument continues, where the copies are digitally created and therefore potentially perfect substitutes for the original.

Copyright holders have long sought to back up their legal control over reproduction with functional control. In the 1970s, copyright owners sought without success to prohibit the sale of videocassette recorders.[69] In the 1980s, copyright owners succeeded in securing a legal prohibition on rental of records or computer software to forestall, it was said, the unauthorized copying that such rental was likely to inspire.[70] In the 1990s, copyright owners and users' groups compromised

---

63. *See id.* § 110(7).

64. *See id.* § 110(5).

65. There is one, sort of. Section 1008 includes a provision that bars infringement suits "based on the noncommercial use by a consumer" of an audio recording device for making "musical recordings." *See* 17 U.S.C. § 1008 (1994). The provision carefully omits any statement that such recordings are not infringement, and was demanded by the consumer electronics industry as a condition for supporting the Audio Home Recording Act. *See infra* note 71 and accompanying text.

66. 17 U.S.C. § 106(1).

67. *See, e.g.,* WHITE PAPER, *supra* note 4, at 64.

68. *See, e.g., U.S. Adherence to the Berne Convention: Hearings Before the Subcomm. on Patents, Copyrights and Trademarks of the Senate Comm. on the Judiciary,* 99th Cong., 1st and 2d Sess. 47 (1987) (testimony of Barbara Ringer, Former Register of Copyrights); Eric Fleischmann, *The Impact of Digital Technology on Copyright Law,* J. PAT. & TRADEMARK OFF. SOC'Y, Jan. 1988, at 5.

69. *See* Sony Corp. of America v. Universal City Studios, Inc., 464 U.S. 417 (1984); *Home Recording of Copyrighted Works: Hearings Before the Subcomm. on Courts, Civil Liberties and the Administration of Justice of the House Judiciary Comm.,* 97th Cong., 2d Sess. 1–2 (1982).

70. Record Rental Amendment of 1984, Pub. L. No. 98-450, 98 Stat. 1727 (1984); Computer Software Rental Amendments of 1990, Pub. L. No. 101-650, 104 Stat. 5134-37 (1990) (codified at 17 U.S.C. § 109 (1994)). *See generally* 2 PATRY, *supra* note 46, at 842–62.

on the adoption of the Audio Home Recording Act,[71] which, for the first time, required that recording devices be technologically equipped to prevent serial copying. The White Paper takes this principle further. It urges a statutory amendment prohibiting any devices or services designed to circumvent technological copy protection.[72]

The Working Group's copy protection proposals have received a lot of attention because they strike a lot of people as monumentally unwise.[73] Devices and services that overcome copy protection have legitimate uses. Copy protection can, after all, block access in situations when the copyright statute would privilege it.[74] At a more basic level, the recommendation seems like an effort to hobble digital technology: to force digital objects to behave as if they were print. I won't repeat all the arguments against the proposal. I instead want to challenge what I take to be its underlying premise: that the right to make copies is fundamental to copyright in any sense other than the historical one.

When the old copyright laws fixed on reproduction as the compensable (or actionable) unit, it wasn't because there was something fundamentally invasive of an author's rights about making a copy of something. Rather, it was because, at the time, copies were easy to find and easy to count, so they were a useful benchmark for deciding when a copyright owner's rights had been unlawfully invaded. Unauthorized reproductions could be prohibited without curtailing the public's opportunities to purchase, read, view, hear or use copyrighted works. They are less useful measures today. Unauthorized copies have become difficult to find and difficult to count. In addition, now that copyright owners' opportunities to exploit their works are as often as not unconnected with the number of reproductions, finding and counting illicit copies is a poor approximation of the copyright owners' injury.

The reasons that copyright owners and the IITF Working Group might have for wanting to treat reproduction as a fundamental copyright right are obvious. By happenstance (at least from the vantage point of 1790, or 1870, or even 1909 or 1976), control over reproduction could potentially allow copyright owners control

---

71. Audio Home Recording Act of 1992, Pub. L. No. 102-563, 106 Stat. 4237 (1992). *See Audio Home Recording Act of 1991: Hearing on H.R. 3204 Before the Subcomm. on Intellectual Property and Judicial Administration of the House Comm. on the Judiciary,* 102d Cong., 2d Sess. (1993).

72. WHITE PAPER, *supra* note 4, at 230–34.

73. *See, e.g.,* Kurtz, *supra* note 11; Pamela Samuelson, *Technological Protection for Copyrighted Works,* 45 EMORY L.J. ___ (forthcoming, 1996); Mark Voorhees, *Keeping Control: Lehman Panel's Report on Net Commerce in Final Phases of Tugs, Pulls, and Faxes,* INFO. LAW ALERT, May 12, 1995; Zimmerman, *supra* note 4, at 408–12.

74. *See* Vault Corp. v. Quaid Software Ltd., 847 F.2d 255 (5th Cir. 1988); Sega Enters. Ltd. v. Accolade, Inc., 977 F.2d 1510 (9th Cir. 1993); Atari Games Corp. v. Nintendo of America, Inc., 975 F.2d 832 (Fed. Cir. 1992); COPYRIGHT AND HOME COPYING, *supra* note 16; Litman, *supra* note 4, at 32 n.21; Samuelson, *supra* note 73.

over every use of digital technology in connection with their protected works.[75] This is not what the Congresses in 1790, 1870, 1909 and 1976 meant to accomplish when they awarded copyright owners exclusive reproduction rights. Printing presses used to be expensive; the photocopy machine was invented during the baby boom. Multiple reproduction was, until very recently, a chiefly commercial act. Pegging authors' compensation to reproduction, therefore, allowed past Congresses to set up a system that encouraged authors to create and disclose new works while ensuring the public's opportunities to read, view or listen to them, learn from them, share them, improve on them and, ultimately, reuse them. Today, making digital reproductions is an unavoidable incident of reading, viewing, listening to, learning from, sharing, improving, and reusing works embodied in digital media. The centrality of copying to use of digital technology is precisely why reproduction is no longer an appropriate way to measure infringement.

As recently as the 1976 general copyright revision, the then-current state of technology permitted Congress to continue its reliance on the exclusive reproduction right by enacting a lot of arcane, hypertechnical rules and exceptions to it, at the behest of all of the stakeholders who argued that they required special treatment.[76] That did not pose major problems because very few people needed to understand what the rules were, and many, if not most of them, could afford to hire lawyers. Unauthorized reproduction was illegal, said the rules, unless you were a "library or archives,"[77] a "transmitting organization entitled to transmit to the public a performance or display of a work,"[78] a "governmental body or other nonprofit organization,"[79] or a "public broadcasting entity";[80] or unless you were advertising "useful articles that have been offered for sale,"[81] "making and distributing phonorecords,"[82] or making pictures of a building "ordinarily visible from a pub-

---

75. See WHITE PAPER, *supra* note 4, at 64-66, 90-95; *see also Sept. 23, 1994 IITF Hearing, supra* note 21, at 6 (testimony of Steven Metalitz, Information Industry Association):

We've got a few advantages into proceeding through the vehicle of the reproduction right. One is it avoids the question of defining or debate over whether a particular distribution of a work has been made to the public, or whether even distribution or a performance has been made to the public. Second, it may obviate the need for an expansion of the definition of publication

. . . .

While the White Paper's characterization of the reproduction right as encompassing any appearance of a work in computer RAM is difficult to defend, *see supra* note 8, that distortion may not have much practical significance. There is little question that the same work saved to a disk cache or backup file is "reproduced" within the literal terms of the statute.

76. *See generally* Litman, *supra* note , at 323-26; Jessica D. Litman, *Copyright, Compromise and Legislative History*, 72 CORNELL L. REV. 857 (1987).

77. *See* 17 U.S.C. § 108 (1994).

78. *Id.* § 112(a).

79. *Id.* § 112(b).

80. *Id.* § 118.

81. *Id.* § 113(c).

82. *Id.* § 115.

lic place."[83] Those entitled to exemptions knew who they were and knew what limitations their privileges entailed.

We no longer live in that kind of world. Both the threat and promise of new technology centers on the ability it gives many, many people to perform the twenty-first century equivalents of printing, reprinting, publishing and vending. Copyright owners all over want the new, improved rules to govern the behavior of all citizens, not just major players in the copyright-affected businesses. And, since anyone who watches citizen behavior carefully to detect copyright violations can easily find enough to fill up her dance card in an afternoon, copyright owners have taken to the argument that citizens must be compelled to obey the rules, by installing technology that makes rule breaking impossible for the casual user and difficult for the expert hacker. Otherwise, they've argued, there is no hope of everyone's obeying the law.

Well of course not. How could they? They don't understand it, and I don't blame them: It isn't a particularly easy set of rules to understand, and even when you understand it, it's very hard to argue that the rules make any sense–or made any sense, for that matter, when they were written. What nobody has tried, or even proposed, is that we either scrap the old set of rules, or declare the general citizenry immune from them, and instead devise a set of rules that, first, preserve some incentives for copyright holders (although not necessarily the precise incentives they currently enjoy); second, make some sense from the viewpoint of individuals; third, are easy to learn; and fourth, seem sensible and just to the people we are asking to obey them.

V

The first task, then, in revising copyright law for the new era requires a very basic choice about the sort of law we want. We can continue to write copyright laws that only copyright lawyers can decipher, and accept that only commercial and institutional actors will have good reason to comply with them, or we can contrive a legal structure that ordinary individuals can learn, understand and even regard as fair. The first alternative will take care of itself: The White Paper comes with its very own legislative proposal that will inspire precisely the sort of log-rolling that has achieved detailed and technical legislation in the past.[84] The second alternative is more difficult. How do we define a copyright law that is short, simple and fair?

If our goal is to write rules that individual members of the public will comply with, we need to begin by asking what the universe looks like from their vantage point. Members of the public, after all, are the folks we want to persuade that copyright is just and good and will promote the progress of science. They are un-

---

83. *Id.* § 120.
84. WHITE PAPER, *supra* note 4, Appendix 1 at 1–12; *see* S. 1284, 104th Cong., 1st Sess. (1995).

likely to think highly of the Working Group's proposal that they need to secure permission for each act of viewing or listening to a work captured in digital form. They are unlikely to appreciate the relentless logic involved in concluding that, while copyright law permits the owner of a copy to transfer that copy freely, the privilege does not extend to any transfer by electronic transmission.[85] They are unlikely to be persuaded that the crucial distinction between lawful and unlawful activity should turn on whether something has been reproduced in the memory of some computer somewhere.

If we are determined to apply the copyright law to the activities of everyone, everywhere, then I suggest that the basic reproductive unit no longer serves our needs, and we should jettison it completely.[86] That proposal is radical: if we stop defining copyright in terms of reproduction, we will have to rethink it completely. Indeed, we will need a new name for it, since *copy*right will no longer describe it. What manner of exclusive right could we devise to replace reproduction as the essential compensable unit?

The public appears to believe that the copyright law incorporates a distinction between commercial and non-commercial behavior.[87] Ask non-lawyers, and they will tell you that making money using other people's works is copyright infringement, while non-commercial uses are all okay (or, at least, okay unless they do terrible things to the commercial market for the work).[88] Now, that has never, ever been the rule but, as rules go, it isn't a bad start. It isn't very far from the way, in practice, the rules actually work out. Non-commercial users rarely get sued and, when they do, tend to have powerful fair use arguments on their side. Moreover, if it is a rule that more people than not would actually obey because it struck them

---

85. *See* WHITE PAPER, *supra* note 4, at 93–94:
Some argue that the first sale doctrine should also apply to transmissions, as long as the transmitter destroys or deletes from his computer the original copy from which the reproduction in the receiving computer was made. The proponents of this view argue that at the completion of the activity, only one copy would exist between the original owner who transmitted the copy and the person who received it–the same number of copies as at the beginning. However, this zero sum gaming analysis misses the point. The question is not whether there exist the same number of copies at the completion of the transaction or not. The question is whether the transaction when viewed as a whole violates one or more of the exclusive rights, and there is no applicable exception from liability. In this case, without any doubt, a reproduction of the work takes place in the receiving computer. To apply the first sale doctrine in such a case would vitiate the reproduction right.

86. My discussion completely omits the immense practical difficulties in getting such a proposal enacted into law, over the presumed antagonism of current copyright stakeholders, and in apparent derogation of our obligations under international copyright treaties. Other copyright professionals who have gone along with my argument thus far are invited to leave the bus at this station.

87. *See* Litman, *supra* note 4, at 35–36.

88. *See, e.g.*, INTELLECTUAL PROPERTY RIGHTS IN AN AGE OF ELECTRONICS AND INFORMATION, *supra* note 16, at 121–22, 209; *see generally*, THE POLICY PLANNING GROUP, YANKELOVICH, SKELLY & WHITE, INC., PUBLIC PERCEPTIONS OF THE "INTELLECTUAL PROPERTY RIGHTS" ISSUE (1985) (OTA Contractor Report).

as just, we would be a long way towards coming up with a copyright law that would actually work. So why not start by recasting copyright as an exclusive right of commercial exploitation? Making money (or trying to) from someone else's work without permission would be infringement, as would large scale interference with the copyright holders' opportunities to do so.[89] That means that we would get rid of our current bundle-of-rights way of thinking about copyright infringement. We would stop asking whether somebody's actions resulted in the creation of "material objects . . . in which a work is fixed by any method now known or later developed,"[90] and ask instead what effect those actions had on the copyright holder's opportunities for commercial exploitation.[91]

Such a standard is easy to articulate and hard to disagree with in principle. The difficulty lies in predicting how it would actually work. So general a rule would necessarily rely on common-law lawmaking for embroidery. One significant drawback of this sort of standard, then, is that it would replace the detailed bright

---

89. Routine free use of educational materials by educational institutions seems like a good example of the sort of noncommercial use that should be classed as "large scale interference" with copyright holders' commercial opportunities. On the other hand, the fact that a particular individual's viewing or copying of a digital work might itself supplant the sale of a license to view or copy if such licenses were legally required should count neither as making money nor as large scale interference with commercial opportunities. For a contrary view, see Jane C. Ginsburg, *Putting Cars on the "Information Superhighway:" Authors, Exploiters and Copyrights in Cyberspace*, 95 COLUM. L. REV. 1466, 1478–79 (1995). Professor Ginsburg argues that because the private copying market has supplanted traditional distribution, even temporary individual copying in cyberspace will impair the copyright owner's rights, although she concedes that fully enforcing those rights may be impractical. *Id.* That copyright holders have recently begun to exploit the market for licenses to make individual copies, however, tells us little about the scope of their entitlement to demand such licenses under current law, and even less about whether a revised law should extend to such claims. *See* Michigan Document Servs. v. Princeton Univ. Press, 74 F.3d 1512, 1523 (6th Cir. 1996) ("It is circular to argue that a use is unfair, and a fee therefore required, on the basis that the publisher is otherwise deprived of a fee."), *vacated en banc*, 74 F.3d 1528 (6th Cir. 1996).

90. 17 U.S.C. § 101 (1994).

91. As an illustration, consider the case of Robert LaMacchia. Mr. LaMacchia was unsuccessfully prosecuted under the wire fraud statute for providing a computer bulletin board where users uploaded and downloaded unauthorized copies of commercially published software. *See* United States v. LaMacchia, 871 F. Supp. 535 (D. Mass. 1994). LaMacchia had no commercial motive and gained no commercial advantage by this activity, but his bulletin board made it possible for some number of people who might otherwise have purchased authorized copies of software to obtain unauthorized copies for free. The White Paper favors amending the copyright law to ensure that people like Robert LaMacchia can be successfully prosecuted for criminal copyright infringement from now on. *See* WHITE PAPER, *supra* note 4, at 228–29. Under the standard I propose, that activity would be infringement if copyright holders demonstrated to the trier of fact that LaMacchia's BBS worked a large scale interference with their marketing opportunities. Merely proving that if such activities were to become widespread they would have potentially devastating marketing effects, on the other hand, would not satisfy the standard.

lines in the current statute with uncertainty.[92] But the bright lines Congress gave us embody at least as much uncertainty, although it is uncertainty of a different sort. The detailed bright lines have evolved, through accident of technological change, into all-inclusive categories of infringers with tiny pock-marks of express exemptions and privileges, and undefined and largely unacknowledged free zones of people-who-are-technically-infringing-but-will-never-get-sued, like your next-door neighbor who duplicates his wife's authorized copy of Windows 95® rather than buying his own from the computer store. The brightness of the current lines is illusory.

Giving copyright holders the sole right to commercially exploit or authorize the commercial exploitation of their works, of course, is a more constrained grant than even the current capacious statutory language, much less the expansive interpretation proffered by the Information Infrastructure Task Force. It removes vexing (if rarely litigated) everyday infringements, like your neighbor's bootleg copy of Windows 95®, from the picture entirely. Is surgery that radical necessary? Certainly not. It would, however, have some significant advantages.

First, to the extent that current constructions of the reproduction right have shown a rapacious tendency, their proponents commonly defend them on the ground that a single isolated unauthorized digital copy can devastate the market for copyrighted works by enabling an endless string of identical illegal copies.[93] Sometimes they explain that a single harmless copy would never give rise to a lawsuit. If that's so, copyright holders lose nothing of value by trading in their reproduction rights for exclusive control over commercial exploitation. If the danger of an unauthorized copy is that it might ripen into a significant burden on the commercial market, then defining that burden as copyright infringement will address the danger without being overinclusive.

Moreover, the common law interpretative process we would necessarily rely on to explicate a general standard unencumbered by all of the detailed exceptions in the current statute is better set up to articulate privileges and limitations of general application than our copyright legislative process has proved to be. While judicial lawmaking may not succeed very well, very often, at arriving at sensible solutions, the process constrains it to try to draw lines that make sense. The public is more likely to accept lines drawn by drafters who are attempting to make sense. And the

---

92. There is a substantial literature on the relative merits of rules and standards. *See, e.g.,* Jonathan Weinberg, *Broadcasting and Speech,* 81 CAL. L. REV. 1103 (1993).

93. *See, e.g.,* Ginsburg, *supra* note 89, at 1467–68; WHITE PAPER, *supra* note 89, at 10, 203. More passionate proponents commonly argue that if the copying is not checked, the incentives for creation will crumble and there will be no copyrighted works to reproduce. *See, e.g., Nov. 18, 1993 IITF Hearing, supra* note 1, at 14–16 (testimony of Steven Metalitz, Information Industry Association). *Cf. Home Recording of Copyrighted Works, supra* note 69, at 142 (testimony of Howard Wayne Oliver, AFTRA) (testifying, in 1982, that if audio and video tape recording were not curtailed, the motion picture and television industries would be crippled and ultimately destroyed).

public's involvement as jurors in drawing these lines just might allow us to incorporate emerging social copyright norms into the rules we apply.

Finally, once we abolish the detailed, specific exemptions in the current law, the industries that have been able to rely on them will need to seek shelter within the same general limitations on which the rest of us depend. It is common for large copyright-intensive businesses to insist that they are *both* copyright owners and copyright users, and that they are therefore interested in a balanced copyright law.[94] They typically fail to mention that, unlike the vast majority of copyright users, and unlike new start-up copyright-affected businesses, they were able to negotiate the enactment of detailed copyright privileges. In most cases, those privileges both gave them what they believed at the time they would need, and also, if they were very clever or very lucky, were drafted with enough specificity to prove unhelpful to new, competing media that might crawl out of the woodwork in the future. Eliminating current stakeholders' structural advantages from the copyright law would do much to restore a more durable balance.

Most copyright owners, of course, would greet my suggestion without enthusiasm. Their reluctance to relinquish theoretical legal control over unauthorized private noncommercial reproduction is likely to be due, in part, to the usefulness of the reproduction right in a digital era as an all-purpose pretext for asserting control over activities never meant to be included within the copyright bundle of rights. The White Paper's analysis illustrates this point neatly. It relies on the reproduction right to vest copyright holders with legal control over "most NII transactions,"[95] and voices no concern that networked digital communications were not even considered by Congress when it enacted the relevant language. The White Paper, moreover, uses the reproduction right to support its argument that adding a new right of transmission to the copyright statute will not expand copyright owners' rights to a meaningful degree.[96] Finally, it is the spectre of mass violations of the fundamental reproduction right that the White Paper raises in support of its proposed prohibition of devices and services that circumvent technological copy protection.[97]

In addition to separating copyright owners from a useful tool for overreaching, abandoning the reproduction right in favor of a right of commercial exploitation

---

94. *See, e.g., Sept. 16, 1994 IITF Hearing, supra* note 1, at 22 (testimony of William Barwell, Times Mirror Company).

95. WHITE PAPER, *supra* note 4, at 64.

96. *Id.* at 216.

Some are of the view that the current language of the Act does not encompass distribution by transmission. They argue that the proposed amendment expands the copyright owner's rights without a concomitant expansion of the limitations on those rights. However, since transmissions of copies already clearly implicate the reproduction right, it is misleading to suggest that the proposed amendment of the distribution right would expand the copyright owner's rights into an arena previously unprotected.

Id.

97. *Id.* at 230–34.

would have the benefit of conforming the law more closely to popular expectations. That would ease enforcement, and make mass education about the benefits of intellectual property law[98] more appealing.

I don't suggest for a minute that limiting copyright's exclusive rights to a general right of dissemination for commercial gain will solve all of the problems I have raised for the public's side of the copyright bargain. Most obviously, copyright holders will rely, as they have in the past, on mechanisms outside of the copyright law to enhance their control over their works. The technological controls that the White Paper is so protective of are one such mechanism.[99] Adhesion contracts purporting to restrict users' rights as part of a license are another.[100] Indeed, one of the most important items on the White Paper's unstated agenda seems to be the reinforcement of industry efforts to find contract law work-arounds for privileges that current copyright law accords to users.[101] Even if the copyright grant is narrowed in scope, the public will need some of *its* rights made explicit. If we find ourselves too fond of the reproduction right, with its venerable pedigree, to abandon it, then explicit public privileges seem essential.

For example, the public has had, under current law, and should have, a right to read. Until recently, this wasn't even questionable. Copyright owners' rights did not extend to reading, listening, or viewing any more than they extended to private performances. A couple of courts adjudicating commercial disputes,[102] though, relied on the exclusive reproduction right as a catch-all right that captures every appearance of any digital work in the memory of a computer. The authors of the White Paper seized on this interpretation as the correct one. They insist that it applies to private individuals as well as commercial actors.[103] This expansion of the

---

98. *See id.* at 201–10. I found the White Paper's treatment of copyright education and its description of its "Copyright Awareness Campaign" somewhat chilling in its single mindedness; it is careful to leave no room for divergent views on the essential goodness of intellectual property ownership.

99. *See* Samuelson, *supra* note 73.

100. *See, e.g.*, Mark A. Lemley, *Intellectual Property and Shrinkwrap Licenses*, 68 S. CAL. L. REV. 1239 (1995); Steven Metalitz, *The National Information Infrastructure*, 13 CARDOZO ARTS & ENT. L.J. 465, 469–72 (1995); Pamela Samuelson, *Legally Speaking: Software Compatibility and the Law*, 38 COMMUNICATIONS OF THE ACM, Aug. 1995, at 15, 20.

101. *See* WHITE PAPER, *supra* note 4, at 49–59. Many industry witnesses at the Working Group's public hearings insisted that contractual and technological means of controlling public access to their wares were far more important tools than copyright law offered. *See, e.g., Nov. 18, 1993 IITF Hearing, supra* note 1, at 27–29 (testimony of Timothy King, John Wiley & Sons); *Sept. 14, 1994 IITF Hearing, supra* note 2, at 20–23 (testimony of Priscilla Walter, Gardner, Carton & Douglas), 44–46 (testimony of Edward Massie, Commerce Clearing House).

102. *See* MAI Sys. Corp. v. Peak Computer, Inc., 991 F.2d 511, 517 (9th Cir. 1993); Triad Sys. Corp. v. Southeastern Express Co., 1994 WL 446049, at *4-5 (N.D. Cal. 1994), *aff'd in relevant part*, 64 F.3d 1330 (9th Cir. 1995), *cert. denied*, 116 S. Ct. 1015 (1996); Advanced Computer Serv. v. MAI Sys. Corp., 845 F. Supp. 356, 362-63 (E.D. Va. 1994).

103. *See* WHITE PAPER, *supra* note 4, at 64–66.

scope of the reproduction right is controversial.[104] Even if it were to become a set-
tled legal rule, recent courts' application of the fair use privilege arguably exempts
most private, temporary copying from the reach of the current statute.[105] Still, fair
use is a troublesome privilege because it requires a hideously expensive trial to
prove that one's actions come within its shelter. So, let's start with an express
ephemeral copying privilege.[106] Commercial broadcasters, after all, have such a
privilege now,[107] and are in a position to do far more damage with theirs; why not
replace it with one for everyone? If temporary copies are an unavoidable incident
of reading, we should extend a privilege to make temporary copies to all.

Moreover, the public has had, under current law, and should have, an affirmative
right to gain access to, extract, use and reuse the ideas, facts, information and other
public domain material embodied in protected works.[108] That affirmative right
should include a privilege to reproduce, adapt, transmit, perform or display so
much of the protected expression as is required in order to gain access to the unprotected elements. Again, both long copyright tradition[109] and current case
law[110] recognize this right, but the White Paper's proposals on technological copy
protection threaten to defeat it. Copyright owners have no legitimate claim to
fence off the public domain material that they have incorporated in their copyrighted works from the public from whom they borrowed it, so why not make the
public's rights to the public domain explicit?

Finally, the remarkable plasticity of digital media has introduced a new sort of
obstacle to public dissemination. Works can be altered, undetectably, and there is
no way for an author to insure that the work being distributed over her name is the
version she wrote. That danger has inspired some representatives of authors and

---

104. *See* Kurtz, *supra* note 11, at 121–22; Litman, *supra* note 4, at 40–43; *see also* Samuelson, *supra*
note 4.

105. *See* Sony Corp. of America v. Universal City Studios, Inc., 464 U.S. 417, 447–51 (1984); Sega
Enters. Ltd. v. Accolade, Inc., 977 F.2d 1510, 1521-23 (9th Cir. 1993); Atari Games Corp. v. Nintendo
of America, Inc., 975 F.2d 832, 843 (Fed. Cir. 1992).

106. *Accord* American Association of Law Libraries et al., *Fair Use in the Electronic Age: Serving
the Public Interest, available online* URL gopher://arl.cni.org:70/00/scomm/copyright/policy/uses
(Working Document, Jan. 18, 1995).

107. *See* 17 U.S.C. § 112 (1994).

108. Some representatives of copyright owners have recently suggested augmenting the rights in the
copyright bundle by adding an exclusive right to *gain access* to protected works and an exclusive right
to *use* protected works. *See, e.g., The 1976 Copyright Act and the Challenge of the Digital Environment*, Harvard Information Infrastructure Project and U.S. Copyright Office Invitational Seminar on
The Future of Copyright Policy, Washington D.C., Dec. 6, 1995 (remarks of Emery Simon, Alliance to
Promote Software Innovation). It's not clear to me what rights or privileges those representatives en-
vision leaving in public hands beyond the right to purchase access and use licenses if the copyright
owners wish to make them available.

109. *E.g.*, Baker v. Selden, 101 U.S. 99 (1879).

110. Sega Enters. Ltd. v. Accolade, Inc., 977 F.2d 1510 (9th Cir. 1992); Atari Games Corp. v. Nintendo of America, Inc., 975 F.2d 832 (Fed. Cir. 1992).

publishers to insist that the law give copyright holders more control over their digital documents, over access to those documents, and over any reproduction or distribution of them. This solution is excessive; there is a more measured alternative. The United States could adopt its own version of an integrity right.[111] Some copyright experts view integrity rights as yet another way that authors exert unwarranted control over the uses of their works, but the right need not be framed that way. Authors have a legitimate concern, and that concern is often shared by the public. Finding the authentic version of whatever document you are seeking can in many cases be vitally important. Moreover, while European integrity rights include the ability to prohibit mutilations and distortions, digital media gives us the opportunity to devise a gentler solution: any adaptation, licensed or not, should be accompanied by a truthful disclaimer and a citation or hypertext link to an unaltered copy of the original. That suffices to safeguard the work's integrity, and protects our cultural heritage, but it gives copyright owners no leverage to restrict access to public domain materials by adding value and claiming copyright protection for the mixture.

CONCLUSION

The most compelling advantage of encouraging copyright industries to work out the details of the copyright law among themselves, before passing the finished product on to a compliant Congress for enactment, has been that it produced copyright laws that the relevant players could live with, because they wrote them. If we intend the law to apply to individual end users' everyday interaction with copyrighted material, however, we will need to take a different approach. Direct negotiation among industry representatives and a few hundred million end-users would be unwieldy (even by copyright legislation standards). Imposing the choices of the current stakeholders on a few hundred million individuals is unlikely to result in rules that the new majority of relevant players find workable. They will not, after all, have written them. There are, moreover, few signs that the entities proposing statutory revision have taken the public's interests very seriously. Instead, they seem determined to see their proposals enacted before they can be the subject of serious public debate.

---

111. "Integrity right" is a term of art for an author's right to object to or prevent mutilation or gross distortions of protected works. *See generally* Edward J. Damich, *The Right of Personality: A Common-Law Basis for the Protection of the Moral Rights of Authors*, 23 GA. L. REV. 1, 15–23 (1988). The Berne Convention, a treaty the United States ratified in 1989, requires its members to protect authors' moral rights, including integrity rights. The United States has relied chiefly on the Lanham Trademark Act, 15 U.S.C. §§ 1051–1127, to fulfill those obligations. In any event, the integrity right I propose is probably more consonant with the Lanham Act's approach to trademark issues than the Copyright Act's approach to authorship rights. For a different spin on integrity rights and the Internet, see Mark A. Lemley, *Rights of Attribution and Integrity in Online Communications*, 1995 J. ONLINE L., art. 2.

If the overwhelming majority of actors regulated by the copyright law are ordinary end-users, it makes no sense to insist that each of them retain copyright counsel in order to fit herself within niches created to suit businesses and institutions, nor is it wise to draw the lines where the representatives of today's current stakeholders insist they would prefer to draw them. Extending the prescriptions and proscriptions of the current copyright law to govern the everyday acts of non-commercial, non-institutional users is a fundamental change. To do so without effecting a drastic shift in the copyright balance will require a comparably fundamental change in the copyright statutory scheme. If we are to devise a copyright law that meets the public's needs, we might most profitably abandon copyright law's traditional reliance on reproduction, and refashion our measure of unlawful use to better incorporate the public's understanding of the copyright bargain.

ACKNOWLEDGMENTS

This chapter was originally published in the *Oregon Law Review*, Vol. 75 (1996), and is reprinted here with permission.

Author Index

Numbers in italic indicate complete bibliographical references; numbers in parentheses indicate footnote citations.

A

Adams, W., 154, 158, *165*
Advisory Council on Information Highway of Canada, 248, *269*
Alliance v. FCC, 224, *228*
Ameritech Corp. v. United States, 220, *228*
Armstrong, M., 45(1), *45,* 56(18), 57(19)
Association of Research Libraries, 238, *244*

B

Baron, D., 23(20), *23*
Barron, J. A., 224, *228*
Baseman, 28(31), *28*
Baumol, W. J., 45(1), *45,* 46(3), 50(9), 84, *95*
BellSouth Corp. v. United States, 218, *228*
Besen, S., 151, *165*
Bohn, R., 79, *94, 95*
Bork, R., 18(9), *19*
Bowen, W. G., 242, *244*
Boyle, J., 273(6), *273*
Braden, R., 119, *121*
Bradford, R., 103, *121*
Brassil, J., 247, *269*
Braun, H., 79, *94, 95*
Braunstein, Y., 249, *269*
Brennan, T. J., 14(1), *14,* 19(10, 12), *19,* 23(20), *23,* 30(41), 28(34), *28,* 29(39), *30,* 31(46), 58(20), *58*
Brock, G., 17(2, 5), *17,* 19(10), 28(32), 154, 158, *165*
Brownlee, N., 103, *121*

C

Carlin v. Mountain States, 223, 224, *228*
Carlton, D., 18(8), *18*
Caruso, D., 279(32), *279*
Casner, S., 118, *121*

Chang, I. Y., 143, 156, *165*
Cheriton, D. R., 118, *121*
Chesapeake & Potomac Tel. Co. v. United States, 217, 218, 219, 220, *228*
Claffy, K., 79, 94, *95*
Clark, D., 79, *95, 96,* 102, 109, 118, 119, *121, 192, 211*
Coate, 17(6), *17*
Cocchi, R., 66(2), *78,* 79, *95,* 98, 118, *121*
Cole, B., 28(32), *28*
Computer Systems Policy Project, 143, *165*
Cook, G., 90, *95*

D

Damich, E. J., 295(111), *295*
Danielsen, K., 91, *95*
David, P. A., 155, *165*
Dee, J., 228, *228*
Deering, S., 112, 114, 118, 119, *121, 123*
Doyle, C., 45(1), *45,* 56(18), 57(19)
Dreyfus, R. C., 282(43), *282*

E

Economides, N., 45(1), *45,* 48(6), *48,* 147, 154, 155, *165*
Edell, R. J., 91, *95, 103, 121*
Egras, H., 45(1), *45*
Eldering, C., 169, 183, *188*
Emory, L. J., 286(73), *286*
Estrin, D., 66(2), *78,* 79, 90, *95, 96,* 98, 114, 115, 116, 118, 119, *121, 122, 123*
Evans, D., 17(5), *17*

F

Farrell, J., 147, 150, *165*
Faulhaber, G., 19(11), *19,* 29(39), *28*

297

Subject Index

DATE DUE

Telecommunications Policy
 Research Conference 1995)

The internet and
 telecommunications policy